# BEEN THERE, DONE THAT...

## GOT THE T-SHIRT!

**MSgt A. A. Bufalo USMC (Ret)**

ISBN 978-0-9845957-8-5

Cover photo courtesy of Cathey Bridges

First Printing – April 2011
Printed in the United States of America

# www.AllAmericanBooks.com

# Been There, Done That... Got the T-Shirt!

# OTHER BOOKS BY ANDY BUFALO

## SWIFT, SILENT & SURROUNDED
*Sea Stories and Politically Incorrect Common Sense*

## THE OLDER WE GET, THE BETTER WE WERE
*MORE Sea Stories and Politically Incorrect Common Sense*

## NOT AS LEAN, NOT AS LEAN, STILL A MARINE
*Even MORE Sea Stories and Politically Incorrect Common Sense*

## EVERY DAY IS A HOLIDAY,
## EVERY MEAL IS A FEAST
*A Fourth Book of Sea Stories and Politically Incorrect Common Sense*

## THE ONLY EASY DAY WAS YESTERDAY
*Fighting the War on Terrorism*

## TO ERR IS HUMAN, TO FORGIVE DIVINE
**However, Neither is Marine Corps Policy**
*A Book of Marine Corps Humor*

## HARD CORPS
*The Legends of the Marine Corps*

## AMBASSADORS IN BLUE
**In Every Clime and Place**
*Marine Security Guards Protecting Our Embassies Around the World*

## SALTY LANGUAGE
*An Unabridged Dictionary of Marine Corps Slang, Terms &Jargon*

## THE LORE OF THE CORPS
*Quotations By, About & For Marines*

*Been There, Done That... Got the T-Shirt!*

4

# IN MEMORY OF

**Sergeant Major Joseph J. Ellis**

2$^{nd}$ Battalion, 4$^{th}$ Marines

KIA February 7, 2007 in Al Anbar Province, Iraq

and

**Lieutenant Colonel Rodney C. "K2" Richardson**
(Retired)

KIA October 4, 2006 near Baghdad, Iraq

**"Peace is that brief glorious moment in history when everybody stands around reloading."** - Thomas Jefferson

# PREFACE

A lot has happened since *Every Day is a Holiday, Every Meal is a Feast* came out in 2006, and while I had fully intended to make that book the last in my "sea stories and politically incorrect common sense" series, there were still a few things which needed to be said. Those serious issues, which are interspersed amongst those which are just plain funny or purely motivational, are the center of what will probably be the final installment. One other factor which influenced my decision was putting together another book gave me an opportunity to honor two of the finest Marines I have ever known in Joe Ellis and Rod Richardson. Men such as those are the benchmark of all we should aspire to be - and I am humbled by having had the privilege to wear the same uniform as they did.

If you have read my first book, *Swift, Silent and Surrounded*, you will know one of the inspirations for me writing in the first place was the work of legendary Major Gene "Dunk" Duncan. Shortly before going to press I learned of his passing, and when that happened we as a Corps lost a big piece of our extended family.

Finally, I would like to thank Cathey Bridges, who provided the cover photo. Cathey is the proud mother of a couple of OIF vets, and in case you haven't figured it out she has a special place in her heart for Marines!

Semper Fi, and never forget.

# TABLE OF CONTENTS

# I LOVE MARINES

John Pressley

*I thought this would be an appropriate first story, considering the photo on the cover. In fact, I think one of those t-shirts should be issued to every American!*

Our flight was some thirteen hours over the Pacific Ocean. We were flying standby, and as such received the last two seats on the plane. Our seats weren't the best. They were on the last row. No room to recline. They were next to the bathrooms, and every now and then there was a "wonderful" smell - but I was not complaining, for in my arms was my new baby girl. We were bringing her home. We were bringing her to freedom.

Born in Deshung, China, Gracie was abandoned. We don't know the circumstances, but are pretty confident it was as a direct result of the "one family, one child" policy - an abomination of individual liberty and decency - but nonetheless she was in my arms and once we landed in San Francisco and passed through customs she would become a U.S. citizen.

After de-boarding the plane Jennifer and I began to struggle our way to baggage claim, and then to customs. We were exhausted from our three week trip and the long flight. We had a tired, cranky and malnourished baby. We had several large pieces of luggage which sat in a pile. We had to

15

move it all through the crowded airport passage to customs, and as I stood there I honestly thought to myself there was no way I could make it.

And then they approached. Four young men with high and tight haircuts, each carrying a large overstuffed green duffle bag and wearing a Marine Corp t-shirt. They came up to me and asked if they could help us with our bags.

Though they each had a large load of their own, they carried some of ours. They helped us get to customs and allowed us to go in line before them. They silently stood watch as Jennifer and my papers were stamped. They stood at attention when the customs agent cleared his throat and asked, "Mr. Pressley. It's my understanding that when I stamp Gracie's passport she will become a citizen of the United States of America. Is that correct?" As I tearfully said yes, those Marines cheered as the custom agent slammed down his stamp and said, "God bless you Gracie, welcome to America!"

Soon after we parted ways, but I will forever remember those Marines. I will never, ever forget that experience, understanding that it is but a small glimpse into the honor that drives Marines. They silently stand watch over our country and stare tyranny in the face and say, "Not on my watch." They stand at attention and answer our country's toughest call. Because of their sacrifice I am free. Because of their sacrifice, I stand and cheer. I love Marines!

# A SWORD FOR A MARINE

**Rachel R. Hartman**

I don't really come from a military family. My maternal grandfather was a sergeant in the 32nd (Red Arrow) Division and a maternal great-uncle is an Airborne veteran, but most of my relatives have been Methodist pastors, teachers, and farmers. I didn't even grow up near a military base, and yet somehow after college I began to collect a number of friends who were veterans. A Marine Master Sergeant, who directed the throwing of birdseed at our wedding. A submariner who offered to play bagpipes as I went down the aisle. A few other Navy enlisted. And one Marine captain, whom I'll simply call Elizabeth.

Elizabeth joined the Marines after college, during the Vietnam period, and her husband served overseas in the Army. One evening many years later a group of us had gotten together, and somehow or another the conversation turned to weddings. Elizabeth was asked if she and her husband had both been in uniform and if both of them had worn their dress swords. Elizabeth's teeth came visibly on edge, and she explained that she had not gotten her sword because during the Vietnam War the male officers in the Marine Corps were considered to have a greater need for the dress weapon. As a female Marine serving stateside, she had never received her Mameluke, a weapon every Marine respects as a symbol of their service. Everyone present agreed that this was annoying and unfortunate, and the conversation moved on to other matters.

The next evening my husband and I went out for supper, and between the sushi and the entrée we reviewed the previous night's conversation, once again coming to the conclusion that this had been rather stinky and how rotten it was that Elizabeth was still angry about it because the problem had never been fixed.

Then my husband looked at me and asked, "Why don't we get her one?" and I said, "Well, I do know this catalog that carries them...."

Over dinner we hashed out the plan. We were confident we could front the cost of the sword, but also felt a big present like this is more fun when it's from a group of people rather than just a couple. Besides, we could easily picture the reaction of Elizabeth's other friends should we shut them out!

The next day I sat down and sent out an e-mail to ten or so of Elizabeth's friends who I thought would be most willing to participate laying out the situation, citing the cost of the sword, and asking who wanted to play along. Naturally, they all did. We discreetly checked with her husband, who was delighted with both the idea and the opportunity to help surprise his wife. A few other names were suggested, so that by the time I got around to ordering the sword the following Monday we had nearly twenty people involved from all parts of the country. This made me confident enough to request FedEx second-day shipping.

Other ideas sprang up. We should have it engraved. We

should get a sword rack. We should buy chocolate. People started sending me money, more than the suggested donation amount, adding, "If you need more, let me know." Others pledged funds as soon as their next payday arrived. We planned to present the sword to Elizabeth at the big anniversary/birthday party at their house in late October. I sent out a request for recommendations for local engravers.

The box from the manufacturer arrived... and thus appeared the first plot complication. They had sent a *Navy* sword. It was a very pretty weapon, but there was no way I could give it to Elizabeth, and I was on the phone to the manufacturer before the bubble wrap had time to hit the floor. Fortunately the nice customer service lady wasted no time in apologizing for their obvious error and promised to send me out a pre-paid UPS label, the better to return the sword.

While the money continued to come in, we started our presentation plans. I claimed the privilege of handing the sword to Elizabeth, but also wanted to make sure the other co-conspirators attending the party would have parts to play as well. Jody, who had known Elizabeth for many years, relayed to us the tradition of paying for a sword (even a gift sword) with a coin so that you do not pay for it in blood, and so Megan (the birthday girl of the weekend) was tapped to present Elizabeth with a suitable coin. Ruta, who had known Elizabeth even longer than Jody, just happened to have a Kennedy half-dollar from 1968 - the same year Elizabeth entered the Marines, which we agreed was a suitable coin.

Megan's mother Beth volunteered a presentation box.

John, an Annapolis graduate and a Navy veteran, told us about military sword holders and how he could get one. Jody started collecting names for an illuminated scroll that would serve as our gift card, which Lee agreed to read in her best field herald voice.

The right sword, the Mameluke, finally arrived and was as magnificent as I hoped it would be. One Thursday, when I knew Elizabeth would not be attending the usual weekly gathering, I brought it over and let David, an experienced swordsman and weapons-maker, check it out. He pronounced it most excellent, and worthy of our favorite Captain. A few other co-conspirators were present and took turns holding it very carefully and making appropriate sounds of awe, mostly with vowels.

The next task was getting it engraved. My day job was getting crazy, especially with an unexpected promotion in the works, so Ruta graciously volunteered to handle the engraving part of the project. Supplied with the names of two engravers, both fortunately close to her home, she set out. The first engraver unfortunately declared himself unworthy of the project, but the second engraver was experienced in such things and quoted a reasonable price for the work.

But the sword wasn't the only item to be engraved. Once John had sent me a spiffy official Marine Corps sword holder, we had to get the nameplate engraved. Fortunately Elizabeth was out of state that week, so I called her husband to get her final rank and years of service. He gave me his best recollection, and then added, "And if that's wrong, it's still pretty close." We agreed Elizabeth would be too polite to mention any slipups, so the sword holder nameplate went off with my husband and me one evening to one of the local malls... and thus began the second plot complication.

Carry-in charge for a non-store item. Okay, I'd figured as much. What I didn't figure on was going back to pick it up, looking at the nameplate taped to the order sheet, frowning, and asking, "Is this the right plate?" The text was correct, but something didn't quite look right to me. We were in a hurry, but when we were in the parking lot heading for the truck I said to my husband, "I really don't think this is the right plate." The proportions looked wrong, and the adhesive backing wasn't as I remembered it. When we got home, I checked the plate against the available space on the holder. I was right - it was wrong. Height wrong, width wrong, proportions wrong - the only thing right about it was the text. I called the engraving store and slapped the manager's pinkies for trying to pass off a substitute as the original. After several sincere apologies from her, we arranged to return to the store to see what arrangements could be made. The sword holder came along for the ride this time. After searching through the plates she had in stock, the manager had to confess they didn't have one which was a precise fit - but they had one which was just a few millimeters off in height which could be cut down to fit. I agreed this was an acceptable solution, and left them to it.

In the meantime, John was not content to supply an official Marine Corps sword holder. Oh, no... he had another surprise in progress, one he shared with me as a deep secret, lest our fellow conspirators be disappointed it if didn't go through. He had a contact on the staff of the Commandant of Marines, and had started the procedure of getting a letter of congratulations written by the Commandant himself, General Jones. Naturally I thanked John for arranging this, and told him I'd reimburse him from the collection funds for FedExing it here so it could arrive in time for the party. FedEx's services were indeed required - it arrived at David's house the morning before the party. John thoughtfully scanned the letter for me and e-mailed it so we would have a backup plan in case FedEx didn't deliver on time. After some thought, I asked Allen to read the letter. John, as I mentioned earlier, is a Navy veteran, and Allen is also a Navy veteran, and since John lived on the East Coast and couldn't make it to the party, and Allen has a nice speaking voice and has known Elizabeth for years, it seemed appropriate that he read it.

While the letter was making its way through the Commandant's office, we began to polish up the presentation choreography. The e-mails were flying, as we knew there would be no opportunity for any rehearsals. My friend the retired Master Sergeant was brought in as a consultant, and suggested we open proceedings with the reading of the traditional Marine Corps Birthday message, adding, "I know I'd like it if I were getting a dress sword." I was never a recruit, and I don't speak NCO, but I figured his suggestion should be ranked somewhere between a request and an order. Frankly, I was thankful. The Marine Corps Birthday message would relieve me of the need to write or improvise an introductory speech, it would set the appropriate tone for the occasion, and it would serve to inform those guests who were unaware of Elizabeth's time of service. Plus reading a speech meant there was a definite beginning and end, which would give my fellow conspirators a better time frame for getting into position. But other matters were not so easily managed. We had originally hoped the Texas weather would be kind enough to permit us to do the presentation outside, a venue preferable to trying to cram all of the guests into the dining room as we conspirators struggled to make our way through the crowd. The week of the anniversary party, however, was one of the wettest weeks of the month, and there was an extremely high probability of rain for the day of the party. I struggled between my wish for a smooth and comfortable presentation, and my Texan's respect for all the rain we could get.

The rain notwithstanding, all of the pieces were finally coming together, and by Friday evening we had the engraved sword, the coin, the sword plaque, the illuminated scroll, and the Commandant's letter.

Saturday morning began earlier than usual. Since Elizabeth was still completely in the dark about her present, she had asked if I would bake some bread to bring to the party. I had no excuse for declining. We had already gone to elaborate lengths to keep the Mameluke a secret, and refusing to bring my usual homemade goodies to the party would've been out of character for me. So I baked the bread, ran down the long checklist of everything we needed to bring, and then my husband and I headed north for the party.

We arrived around one o'clock, and Elizabeth happened

to be out giving some of the guests a walking tour of their eighty acres - which gave me a chance to confer with my fellow conspirators. A few minor adjustments were made to the presentation plan, and we set three o'clock as zero hour.

There was one minor plot snag. David had very carefully put the Commandant's letter in his van so he would not forget it, and then forgot that he had agreed to loan his van to his friend Michelle so she could do some moving. Luckily Michelle was brought up to speed, and managed to make it to the party before zero hour. Allen had prepared for this by printing off the e-mailed text I'd sent him Friday night, but was happy to have the official copy in the official folder. Zero hour approached. We warned some un-involved guests to stick around for the cake-cutting, which ended up being done inside due to the predicted dreary weather. We all sang "Happy Anniversary to You" and "Happy Birthday to You," and then as soon as some more cake had been passed around I checked with Elizabeth's husband, and he agreed that now was as good a time as any.

So I called for everyone's attention, explained there was a birthday celebration coming up the following month that meant a lot to Elizabeth, and said that I knew we were jumping the gun a bit - but since everybody was here, I hoped Elizabeth wouldn't mind the anticipation. I remembered to say, "I quote," and started reading the birthday message.

"On November 10, 1775, a Corps of Marines was created by a resolution of Continental Congress. Since that date many thousand men have borne the name 'Marine.' In memory of them it is fitting that we who are Marines should commemorate the birthday of our Corps by calling to mind the glories of its long and illustrious history.

The record of our Corps is one which will bear comparison with that of the most famous military organizations in the world's history. During ninety of the 146 years of its existence the Marine Corps has been in action against the Nation's foes. From the Battle of Trenton to the Argonne, Marines have won foremost honors in war, and in the long eras of tranquility at home, generation after generation of Marines have grown gray in war in both hemispheres and in every corner of the seven seas, that our country and its citizens might enjoy peace and security.

In every battle and skirmish since the birth of our Corps, Marines have acquitted themselves with the greatest distinction, winning new honors on each occasion until the term 'Marine' has come to signify all that is highest in military efficiency and soldierly virtue.

This high name of distinction and soldierly repute we who are Marines today have received from those who preceded us in the Corps. With it we have also received from them the eternal spirit which has animated our Corps from generation to generation and has been the distinguishing mark of the Marines in every age. So long as that spirit continues to flourish Marines will be found equal to every emergency in the future as they have been in the past, and the men of our Nation will regard us as worthy successors to the long line of illustrious men who have served as 'Soldiers of the Sea' since the founding of the Corps."

Unsurprisingly, Elizabeth stopped cutting more cake and just about came to attention as soon as she heard "On November 10, 1775, a Corps of Marines was created..." Like any good Marine, she raised a loud "Semper Fi!" at the conclusion of my reading, then like any good hostess resumed cutting the cake. When I signaled Megan to give her the coin, Elizabeth recognized the significance of the date. While she was admiring the coin (thinking that was all there was to everything) we got the scroll to Lee, who started reading. At the same time, I got the sword and held it behind my back until Lee got to my cue. It was actually easier to hide this than I thought because it was so crowded in there, and I was wearing a rather bulky sweater - though I regretted the sweater later as it got muggier.

For the record, Elizabeth's expression when she realized what we'd done was everything we could've wanted. Shock, delight, awe... and a rather unladylike exclamation when I handed her the sword. She barely got it unwrapped as Lee finished reading. Absolute joy, and definite blurry eyes as she read the inscription on the scabbard: "With thanks for your service."

At this point, Elizabeth thought it was over. She had no idea. Allen started reading the Commandant's letter, and if I thought Elizabeth had come to attention during the birthday message, that was nothing compared to how she reacted to hearing the words "One of the great pleasures of serving as Commandant...." She just about started crying again as Allen read. Allen later told me he quit trying to look at her - I'd had a similar problem during the birthday message. Both

Allen and I started out using the good public-speaker technique of lifting your gaze to your audience every sentence or so, but ended up staring at the page because we were in danger of breaking up.

Elizabeth's reaction to the Commandant's letter was, "How the hell did you manage this?" I explained that John has contacts, and she shook her head in amazement. Allen came over and handed her the folder, coming to attention and snapping off a perfect salute, which was amazing considering how crowded it was. I was standing between them, and unfortunately couldn't get out of the way any more than I already was. But there was enough room that Allen didn't whack me with his hand, and Elizabeth was able to return it just as gravely.

Julia handed over the gift-wrapped sword holder, and Beth signaled the start of the Marines' Hymn. "From the Halls of Montezuma" rattled the walls, with Elizabeth joining in with plenty of vigor, even as she was tearing the wrapping paper off. She grew blurry-eyed when she saw the nameplate engraved with her final rank and years of service, but came back in time to finish out the verse.

Presentation over, we headed to the living room so Elizabeth could sit down and admire her present. I told her how the Commandant's letter had been personally signed, not by auto-pen, and she got the shivers. There was much admiring of the Mameluke, the plaque, and somewhere in there I ran out to our truck and presented her with the chocolate, explaining that we'd had a surplus and had turned it into Godiva dark chocolate. She approved.

Now here's a final plot twist. Remember the conversation I described earlier, and how it was the start of this whole grand conspiracy? It turned out that afterwards Elizabeth had reflected on how she was still cross about not getting her sword, and had actually started shopping for one, but then she decided she should just let it be, and if she still wanted one in a year, she could get it then.

Naturally she's happy it worked out the way it did, and I'm certainly am too. As I said to Elizabeth, we did this "out of love, and because it needed doing." The best motivations in the world!

# HE FOUGHT FOR US
## *But He's Not Worthy*

**Neal Boortz**

*You probably know Neal Boortz is a nationally syndicated radio host, but what you probably don't know is his father, Lieutenant Colonel Neal Boortz, Sr., served as a Marine Corps pilot is WWII, Korea and Vietnam.*

His name was Gregory Boyington. Some called him "Pappy," others "Gramps."He served as a combat pilot in World War II with the 1st Squadron, American Volunteer Group, which was better known as the Flying Tigers of China. Boyington later served as a fighter pilot in the Marine Corps, and commanded Marine Fighting Squadron 214. Perhaps you've heard of them. It was called the Black Sheep Squadron, and was later featured in a TV series called *Baa Baa, Black Sheep.* Boyington shot down twenty-eight Japanese aircraft while serving in the Pacific, was later shot down himself, and spent twenty months in a Japanese POW camp. For those of you who aren't up to par on World War II history, Japanese POW camps were not happy places. Torture was common… and we mean *real* torture, not stripping them naked and taking snapshots. After the war Pappy Boyington was awarded the Navy Cross and the Medal of Honor. He died in 1988, and you can visit his grave in Arlington National Cemetery.

Education? Oh yes! Almost forgot! Pappy Boyington was a graduate of the University of Washington. In fact just recently the idea of erecting a memorial to this Medal of Honor recipient on the University campus made its way to the student Senate. Here you have an alumnus who served in World War II, was captured and held prisoner, and was later awarded the Medal of Honor and Navy Cross - so perhaps some sort of monument would be a good idea!

Well, not to Jill Edwards. Thanks to the folks at *WorldNetDaily* you can see a copy of the minutes of a meeting of the student Senate at the University of Washington. Under "old business" there was discussion of a resolution calling for a tribute to Pappy Boyington. Student Senate member Jill Edwards immediately moved to table the resolution. She wanted other issues to be considered. Another member said the issue was at the top of the agenda, and should be dealt with. Jill's motion failed, but she wasn't through. There was then some discussion on why Andrew Everett, another student Senate member, wanted the memorial. Everett responded that Colonel Boyington "had many of the qualities the University of Washington hoped to produce in its students." Well, I guess that might be true, if leadership and courage are considered to be good qualities. Anyway, that's when Jill Edwards spoke up and showed her true colors. She questioned whether it was appropriate to honor a person who killed other people. Then she said a member of the Marine Corps was not an example of the sort of person the University of Washington wanted to produce.

Shall I repeat that? Jill Edwards, a Junior in Mathematics

at the University of Washington, said a U.S. Marine is not - that's right, NOT - an example of the sort of person the University of Washington wants to produce. Let that sink in. To all of you men and women out there who have served with pride in the United States Marine Corps... to those of you who fought in World War II, Korea, Vietnam and the Middle East... Jill Edwards, student Senate member, thinks you are unworthy to be graduates of the University of Washington. My father was a Marine. He's buried in the National Cemetery at Fort Sam Houston in San Antonio, Texas beneath a grave market that reads, "Neal A. Boortz, Sr., LtCol, USMC. World War II, Korea, Vietnam."

I think Jill Edwards is an ignorant fool. I would submit that Jill Edwards was an embarrassment to the University of Washington. With her mathematics degree and her leftist outlook on life, my guess is she'll end up being a teacher in a government school. Oh goody.

By the way... there was at least one more moonbat in the University Senate. Her name is Ashley Miller. Ashley said there were already enough monuments at U of W commemorating "rich white men." Well, I guess you have to get that wealth-envy stuff in there somewhere.

Don't you just love these young people? They're so much fun to watch during those magic years when they know everything and have all the answers to every problem facing mankind. As I have said before, we should take one hundred volunteers from university student Senates across the country - and let's make sure Jill Edwards is one of them - and give them a country to run for four years. Haiti would do just fine.

# SUPERMAN

*Joe Ellis was a friend of mine. We were both Force Recon Communicators, and our paths crossed in some unusual ways over the years. On one memorable occasion we were going through a school in 29 Palms, and I mentioned I was going to San Juan Capistrano during the coming weekend to run in a 10K. He though that sounded like fun, decided to come along, and ended up winning the race! On another occasion I was in "deep kimchi" while on MSG duty in the Congo (see my book "Swift, Silent and Surrounded") and the simple fact that I knew Joe, who had been in the investigating officer's platoon during the Gulf War, got me pretty much off the hook. He was a Marine's Marine.*

His daughter Rachael called him "Superman." Everyone else called him Sergeant Major. If you're ever looking for a textbook definition of a Marine, look no further than Joe Ellis. There is very little which can be added to what has already been written and said about the Sergeant Major, and the actions he took to protect his Marines tell far more about the man than I ever could hope to.

On February 7, 2007 the Sergeant Major was commanding a Marine checkpoint near a crowded place in Iraq. He saw a man walking toward his checkpoint and correctly identified him as a suicide bomber. Out of options, the Sergeant Major did the only thing he could to protect his men. He put himself between the bomber and his Marines. The suicide bomber quickly detonated himself, and Joe Ellis was killed instantly.

I don't know how well Sergeant Major Ellis knew Corporal David Emery, Jr. Maybe he knew him well, and maybe he didn't. Regardless of what their relationship was, the Sergeant Major absorbed just enough of the blast to spare the life of the Corporal.

Emery - DJ to his friends - was on his second tour in Iraq, and was planning to leave the Marine Corps that June. His mother remembers saying goodbye to her son before he deployed in the fall. "He had a bad feeling," she said. "He said, 'Mom, something doesn't feel right this time.'" Then, a few months after he deployed, DJ found out that his new wife Leslie was pregnant. He wrote to his mother, and made her promise to help take care of the baby if anything happened to him.

On February 7th, just one month after his unit was extended in Iraq, DJ was standing near a checkpoint in Al Anbar Province when his battalion Sergeant Major - Joe Ellis - saw a suspicious person approaching. Ellis put himself between his Marines and the suicide bomber just as the man opened his jacket, spread his arms wide, and detonated his explosives.

Ellis was killed instantly… but his body absorbed enough of the blast to give DJ a chance at survival. His legs and left arm were shattered, and he suffered extreme trauma to his abdomen. DJ remembers lying on the ground after the blast, unable to see or feel his legs. He never even saw the bomber.

Corporal Emery spent days in a combat hospital in Baghdad before he was stable enough to move to Landstuhl. DJ's mother Connie and young wife Leslie were told initially that he just had shrapnel wounds to the legs. They waited for more than two days before they got the call to go to Germany immediately because DJ might not make it. Today, Connie and Leslie both just shake their heads when asked to describe how he looked when they arrived in Germany. "He was swelled up bigger than all of us together," Connie said, adding, "and his eyes were swelled open."

Over the next few weeks DJ died on the operating table six times and received more than three hundred units of blood. He had so many transfusions that his blood type actually changed to O-Positive. The doctors in Germany and then Bethesda completely re-built his legs, but the infection became too strong and nearly two months after the attack they amputated one of his legs. Two days later, they took the other leg.

Then, as DJ lay unconscious on the sixth floor of Bethesda Naval Medical Center, his wife Leslie was admitted to the third floor - the maternity ward - and on April 21, 2007 their daughter Carlee was born. DJ was still in and out of consciousness when his mother came in to tell him the news. He just opened his eyes and said, "Okay." Two days later he really awoke for the first time and realized his legs were gone. He cried for awhile when he found out. "It sucked," he says now.

As this is written DJ Emery, his wife, their new baby, and his mother Connie are all living together in a cramped room at Walter Reed Army Medical Center. DJ says he does not regret joining the Marine Corps or serving in Iraq, and when he was asked what was getting him through this difficult time he choked up and said softly, "Family."

DJ has months of rehabilitation ahead of him. He winces in pain with every exercise, but in the three hours of therapy that we watched he never once complained. Corporal Emery was recently promoted to Sergeant and personally received a Purple Heart from President Bush a few weeks ago - a rare honor for a wounded Marine.

While staying in Bethesda near his son DJ's father, David Emery, attended a funeral at Arlington National Cemetery for the Marine he believes saved his son's life.

Our prayers are with Corporal Emery and his young family. There are probably many reasons why Sergeant Major Ellis put himself between the terrorist and Corporal Emery - but first and foremost among them is the fact he was one of his Marines.

Another reason, perhaps, is so that another little girl can look at her father and call *him* "Superman."

Godspeed Joe. Greater love hath no man.

# FIRST MAN ON THE MOON

*My Father went to work for Grumman Aircraft after returning from WWII, and one of the projects he worked on was the Lunar Module for the Apollo Program. We lived near Kennedy Space Center during the late 1960's, and I even got to see Apollo 11 blast off from our backyard!*

Every schoolboy knows that Neil Armstrong was the first man to set foot on the moon, but most people don't realize he was selected to have that distinction ahead of his crewmate - Air Force Colonel Edwin "Buzz" Aldrin - in large part because he was a civilian. NASA was afraid the sight of a military man planting the flag on the lunar surface would give rise to charges the United States was "militarizing" space, so they went ahead and picked a wholesome, All American, non-uniformed astronaut to lead the way.

At some point Armstrong must have realized the entire world would be listening when he became the first human being to speak from the surface of another celestial body, and that whatever words he uttered would be memorialized for all time. It must have been a daunting task for him to come up with a phrase which was simple, yet memorable, and it's not unreasonable to assume he went through many different drafts before settling on the famous, "One small step for a man, a giant leap for mankind…"

There is also a much circulated urban legend which asserts

# Been There, Done That... Got the T-Shirt!

Armstrong, just before he re-entered the lunar lander, made the remark "Good luck, Mr. Gorsky." Many people at NASA thought it was a casual remark concerning some rival Soviet Cosmonaut, however upon checking there was no Gorsky in either the Russian or American space programs. Over the years many people questioned Armstrong as to what the statement meant, but Armstrong always just smiled. Then one day many years later, while answering questions following a speech, a reporter brought it up again - and this time he finally responded, as Mr. Gorsky had died and Armstrong felt he could finally answer.

The story goes that when he was a kid he was playing baseball with a friend in the backyard, and the friend hit a fly ball which landed in his neighbor's yard by the bedroom windows. His neighbors were Mr. and Mrs. Gorsky, and as he leaned down to pick up the ball young Armstrong heard Mrs. Gorsky shouting at Mr. Gorsky. "Sex? You want sex? You'll get sex when the kid next door walks on the moon!"

It just goes to show there is always hope... but I digress. Getting back to the matter of the first man on the moon, I firmly believe things would have been much simpler if it had been a Marine. The obvious choice would have been John Glenn, who was already a national hero after flying jets in combat, setting a cross country speed record, and becoming the first American to orbit the earth. Just imagine him descending the ladder, bounding to the surface, and moments later keying his radio microphone to utter the only words a Marine - *any* Marine - would or could say on such an occasion: "The Marines have landed!"

37

# MEMORIAL DAZE

Kate O'Beirne

*Some of the battles in Marine Corps history have been fought with words rather than weapons, and this is a perfect example. The one constant is the outcome - as usual, the Corps emerged victorious. This article was written in 1999, and this particular battle went on until 2002, when "the Air Force Memorial Foundation Board of Trustees considered it to be in the best interest of all parties to work with Congress and relocate to another site." Finally dedicated in 2006, the Air Force Memorial site was moved to Fort Meyer, on the opposite (south) side of Arlington Cemetery - which, appropriately enough, is much closer to the Army/Navy Country Club. That should make the zoomies happy!*

The Air Force and the Marines at war.

Last fall, with a threatened government shutdown hours away, congressional leaders and senior White House officials huddled to negotiate the crucial final details of more than four hundred billion in spending, but negotiators put aside disagreements over major issues like abortion and labor regulations to handle a dispute just as hot - the duel between the Air Force and the Marine Corps over the site of a memorial.

In 1993 Congress approved a request from the Air Force to build a monument, with private funds, on an unspecified site, and the Air Force Memorial Foundation began a fundraising drive led by Joe Coors Jr. and boosted by defense contractors. After reviewing proposed locations around Washington, the foundation settled on a lovely spot, a quiet twenty-five acre hillside near Arlington Cemetery in Virginia which enjoys a commanding view of the capital's monuments across the Potomac River. Since 1954 this same hillside, known as Arlington Ridge, has also been home to the Iwo Jima Memorial.

The Iwo Jima Memorial commemorates the twenty-six thousand Marine casualties suffered in the battle for that remote Pacific island in 1945. The sculpture is based on a famous photograph of a band of Marines raising the American flag on top of Mount Suribachi in the midst of fierce fighting, and over the past forty-five years the memorial has come to represent the sacrifice of all fallen Marines - and the Corps' veterans are determined to protect their hill from the threatened Air Force invasion.

Understandably so. There is more at stake in the fight than inter-service bragging rights. There is indeed a touch of "Kulturkampf" behind the plans for a slick new modern memorial near the Iwo Jima sculpture, which Washington art poobah J. Carter Brown once dismissed as :kitsch.'" And there certainly is a huge helping of inter-service envy as the Air Force - formed in 1947 - seeks by association some of the glory of the tradition-soaked, 224-year-old Marines.

The services have drafted veterans in Congress, bureaucrats, lawyers, and defense firms in a struggle which threatens to sour the branches' relationship for years to come. In official backing, the Air Force so far has the advantage. The National Park Service, the National Capital Planning Commission, and the Commission of Fine Arts have all endorsed the proposed two-acre site and the memorial's design - which calls for a fifty-foot-tall aluminum, origami-like structure that would include an underground visitors center.

Supporters of the Air Force memorial argue the Marine Corps tacitly approved the proposed location in 1994 when Corps Commandant Carl E. Mundy Jr. refrained from objecting. For his part, Mundy says he wasn't informed just how intrusive the Air Force memorial would be. In any case the Marines have now formally objected to the plans, and more than two dozen former Marine generals, including five former Commandants, have appealed to the secretary of the interior to preserve the current status of the "serene and contemplative park" on Arlington Ridge.

Complaints about the proposed memorial from Marines in the field have reached the Pentagon, and retired Marine Corps Lieutenant General Charles Cooper, chairman of the Iwo Jima Preservation Committee, recently told his supporters, "At this time, we need to avoid looking back, avoid any self-recrimination about whether or not our listening posts were on full alert. The fact is the enemy has penetrated our wire, and our mission is to restore the integrity of our position."

How deep run the feelings? Former Navy secretary James Webb - the son of an Air Force veteran - sees a lack of gratitude for the Marine Corps he himself served. "The very mission in the battle of Iwo Jima," he wrote, "carried out at the cost of a thousand dead Marines for every square mile of territory taken, was to eliminate enemy fighter attacks on Air Force bombers passing overhead and to provide emergency runways for Air Force pilots who had flown in harm's way."

Retired Marine Colonel Dave Severance, who commanded the troops who raised the flag on Iwo Jima, calls the proposed monument a historic insult. "I am convinced we Marines have done enough for the Army Air Corps/U.S. Air Force without having to share our Marine Corps War Memorial park with them."

In response, the Air Force invokes the 52,173 combat deaths suffered by the Army Air Forces in World War II who "deserve a similar respect," as one retired Air Force general put it. Another retired general, Chuck Link, president of the Air Force Memorial Foundation, argues that the Marines are reacting to an "imagined slight" because "there is no way our memorial interferes with their statue." The Air Force memorial would be five hundred feet away from the Marines' monument, and shielded from it by a stand of evergreens.

But Marine supporters don't care how separate the new monument is. The Friends of Iwo Jima and Marine veterans have recently retained a top- notch law firm to battle the National Park Service over the environmental impact of the proposed memorial. "Marines never quit. We will fight it out to the end, bitter or otherwise," declares former Congressman Gerry Solomon, also a former Marine.

Last year during the budget negotiations, it was Solomon

who negotiated an agreement to consider moving the proposed Air Force site, only to have Senate Appropriations Committee chairman and former fighter pilot Ted Stevens nix the arrangement.

If construction of the Air Force Memorial is stalled until the end of next year, congressional approval of any monument would have to be reaffirmed, and the battlefield would again become Capitol Hill. Solomon's fellow former Marines, including senators Pat Roberts, Conrad Burns and Craig Thomas, will continue his fight, while Stevens, with the support of minority leader Tom Daschle - an Air Force veteran - will continue to front for the Air Force.

The Marines are willing to help the Air Force win quick approval for an alternative location near Arlington Cemetery, which would indeed be suitable. Through forty-five years of quiet prayers, sunset parades and wreath-laying ceremonies, the Marines have an emotional and traditional claim on Arlington Ridge that ought to be honored. A new Air Force Memorial right next door would be a monument, among other things, to disrespect. Surely the Air Force can commemorate its fallen just as well elsewhere and, rather than bring the Marines to their knees, content itself for now with the defeat of a far more fitting adversary - Slobodan Milosevic.

**This story originally appeared in the *National Review* on June 28, 1999.**

# IRONMAN AND IRON MIKE

*That last dustup with the USAF was unfortunate, but even so our disagreements with the Army go WAY back to the days before the Air Force even existed... and even then we usually came out smelling like a rose.*

Once World War I was over, General "Blackjack" Pershing commissioned French sculptor Charles Raphael Peyre to create a bronze statue to commemorate the U.S. Army Doughboys' service in WWI. General Pershing told his staff to furnish a model to pose for the French sculptor for his commemorative statue, but apparently not too much guidance was given - because the individual assigned to pose for the statue was a Marine Private named Carl J. Millard of the 75th Company, 6th Marine Regiment, who had been twice wounded at Belleau Wood.

The Frenchman, having no intramural rivalries in his psyche, modeled the Marine Private in his entirety - complete with the Marine Corps Emblem on his helmet! When General Pershing saw the finished statue he refused to accept the Frenchman's work of art – he was, in a word, outraged!

One of the witnesses was a young private named Bill "Ironman" Lee, the legendary three-time Navy Cross recipient who later fought alongside Chesty Puller in Nicaragua and rose to the rank of Colonel. "I was standing about forty feet away. They unveiled the statue... General

Pershing walked to the front of the podium and looks down to see 'Iron Mike' with a Marine emblem on his tin hat... He took a good look, did an about face, and he and his staff marched off the podium."

General Douglas MacArthur was also quite upset, and continued to hold a grudge even after he fled Corregidor in the early days of WWII. When safely ensconced in Australia, "Dugout Doug" immediately wrote each unit left behind on the "Rock" up for a Presidential Unit Citation - all except one, the 4th Regiment of Marines - which was ironic (perhaps "criminal" is a better word), since the Corregidor garrison was ordered to surrender by Army General Jonathan Wainwright, despite the strenuous objections of the 4th Marines' commanding officer, Colonel Sam Howard. When his oversight was pointed out to him, MacArthur ground his teeth and made a statement to the effect that the Marines had garnered unfair publicity in WWI, and he was not going to add to their fame and glory in "THIS" war! Equally baffling was MacArthur being awarded the Medal of Honor for his defense of the Philippines - despite the fact he was evacuated to Australia by PT boat while his command was left behind to fight on. It wasn't until the Inchon Landing in Korea that he finally forgave the Marines their earlier "indiscretions" – after they had once again pulled the Army's chestnuts out of the fire. From that time on they became "his" Marines, and apparently all was forgiven - some thirty-two years after the fact.

Help was in the wings concerning the now orphaned statue with the Marine Corps emblem however, as Smedley

Butler saw it and fell in love. The Chief Paymaster for U.S. Marines in France, Major D.B. Wills, suggested to Major General Commandant George Barnett on March 24, 1919 that the Corps purchase the statue for use as a memorial to the Marines who had given their lives in World War I. The asking price for the statue was fifty thousand francs, so Butler took up a collection from all the Marines in the AEF - one dollar per man - bought the statue from the Frenchman, and shipped the artwork back to the United States where it was placed in front of the old Headquarters Building of the Marine Corps Base at Quantico. Today the original statue - whose official name is actually "Crusading for Right" - remains in front of Butler Hall, which was named after the two-time Medal of Honor recipient and is now home of the Marine Corps Training and Education Command.

# FOR THE WOUNDED
## *No Miracle is Small*

David Zucchino

*This story, and many others like it, are important because they show the need for Americans to support the Fisher House and Semper Fi Injured Marine Fund charities – without which, many of these families could not be by their Marines' side when they need them most.*

The Ryan family stood vigil, gathered around a hospital bed in Building Ten, Ward Five East - a surgical ward at the National Naval Medical Center at Bethesda. Before them lay Corporal Eddie Ryan, silent and pale, a grievous bullet wound in his brain and a feeding tube in his belly, straight through the "N" in a blue tattoo that spelled "RYAN."

Angela Ryan stroked her son's fine hair. Christopher Ryan squeezed his boy's hand. Nineteen-year-old Felicia Ryan looked into her brother's eyes, her hand on a Bible which was resting against his left leg.

The news was not good. Eddie's neurosurgeon, Robert Rosenbaum, had told the family the young Marine's frontal lobes had been terribly damaged by the bullet that tore into his skull during a firefight in western Iraq. It was quite possible Eddie, who was only twenty-one, would never fully regain consciousness or return to what the doctor called "full cognitive activity."

46

Christopher stared at his son's smooth face and spoke. "We need a miracle. Eddie's going to be our miracle Marine. We're praying that God gives us this miracle because my son is a great American."

Across the hall on that same day Marine Corporal Bryan Trusty sat up in bed, wolfing down a chicken dinner on a hospital tray. His father sat at his bedside, amazed his son was eating, talking, and even laughing.

Earlier that month a hot shard of shrapnel had ripped a hole beneath Bryan's left eye, pierced the length of his brain and lodged against his brain stem. He survived emergency surgery in Baghdad, but went into cardiac arrest on the medevac flight to the U.S. - and his doctors did not expect him to live.

Now Bryan, who turned twenty-one while in the intensive care unit four weeks earlier, was about to be discharged for outpatient therapy - with shrapnel still in his brain and arm, and a distinct memory of all that had befallen him. He is able to walk and speak normally.

"I call him my miracle child," his father said while watching him eat.

The number of service members wounded in Iraq has surged, with half of them injured so badly they cannot return to duty. Many of the most critical cases end up here at the National Naval Medical Center, which was established in the early days of World War II.

On the worst nights at the Bethesda complex, ambulances and casualty buses deliver up to a hundred wounded Marines and sailors from Iraq. Most of them are young, and suffering from the devastating damage inflicted by explosives, bullets and shrapnel.

Some, like Bryan Trusty, stay only a few weeks. Others, like Eddie Ryan, stay longer. In every case they are surrounded by attentive nurses and skilled surgeons, and by loved ones who cling to hope and share an ordeal which can be both traumatic and uplifting - their lives in turmoil, and forever altered.

If not for Eddie's tattoos, Angela Ryan would not have recognized her own son after she and her husband flew to see him at a military hospital in Germany. His face and body were grotesquely swollen. Before he was wounded Eddie was lean and fit, six feet tall and one hundred and ninety-five pounds, but he had ballooned to two fifty because of severe swelling and fluid accumulation caused by injuries.

Eddie is a sniper, one of the Marine Corps' elite. He signed up straight out of high school, was sent to Iraq, and was on his second tour there when an enemy bullet pierced his brain.

Since he arrived here last month, his mother has not left his side. She sleeps in his room or down the hall in the visitors' lounge. His father and sister have left the hospital once, to drive Eddie's beloved black Toyota Tacoma pickup from a friend's house in Virginia to the family home in Ellenville, N.Y.

The Ryans have taken leaves from their jobs - Christopher as a heavy equipment operator, and Angela as a school lunchroom monitor. Felicia has left community college, as well as a job at an outdoor supply store.

"Wherever Eddie is, that's our life now," Felicia said.

The Bible resting on the hospital bed contained a photo of Eddie in his Marine dress uniform, looking handsome and fearless. Placed between his feet were an embroidered Marine Corps logo and a photo of Eddie and his family the day they picked him up at Camp Lejeune, N.C. when he returned safely from his first tour in Iraq.

Much of the time, Eddie's eyes were open. He breathed on his own, but he did not speak. He was shirtless, and his tattoos were on display. His parents had not approved of them - for a while, Eddie wore long-sleeved shirts to hide the ink - but now the Ryans found comfort and inspiration in the body markings.

On Eddie's abdomen is the RYAN tattoo. On his right arm is a tattoo of hands in prayer and the Marine Corps logo. On his left arm is an American flag, and the words "Land of the free because of the brave."

His parents read to him from the Bible, squeezing his hands as they prayed. His father read him a favorite passage, Psalm 50, Verse 15: "And call upon me in the day of trouble. I will deliver thee and thou shall glorify me...."

"Eddie's faith is very important to him," Angela said, careful as always to speak of her son in the present tense.

In the evenings, she sang to him. He loved being sung to at night as a boy, and often would fall asleep to "Jesus Loves Me, This I Know." She sang him the hymn, softly.

Sometimes Angela sang a favorite song titled "Show Me Your Ways," which includes the lines, "Show me your ways that I may talk with you... to live with the touch of your hand, stronger each day, show me the way."

"That's my prayer," she said, "to walk with him and talk with him again."

Rosenbaum, the neurosurgeon, was candid with the Ryans about their son's prospects. "From the beginning, I made it very clear to Eddie's mom and dad that if we were successful in keeping him alive after the initial swelling, they'd be taking home a human form, but I did not think they'd be able to take home Eddie as they remembered him. He will not remember his dog or his best friend or his room at home. If a miracle of miracles happens and he wakes up and begins to interact, they'll have a person they've never met before. And unfortunately, that is the extreme best scenario."

The bullet or bullet fragments that penetrated Eddie's brain created a percussive wave that produced a temporary cavity and caused severe bilateral frontal lobe injuries, Rosenbaum said. Portions of his skull were destroyed on impact, while others were removed by surgeons in Iraq to relieve brain swelling.

"The doctor told us that Eddie lost two-thirds of his frontal lobes," Christopher said. "And the frontal lobes are what make Eddie, Eddie."

However, Rosenbaum said, Eddie's youth and superb physical condition can improve his degree of recovery. He's a fitness fanatic, and an amateur boxer. While home on leave, he would jog in his combat boots while lugging a backpack loaded with rocks.

Twice a day, Rosenbaum checked Eddie for a response to a stimulus - a pinch, the squeeze of a hand. He said this month that he had not detected anything more than purely reflexive responses, but he promised the Ryans that he would devote "200% effort" to their son.

"Not only Eddie, but Eddie's parents have made the sacrifices that the country asked of them," Rosenbaum said. "It's the responsibility of those of us here to do our very best for them."

In Eddie's room, Felicia tried gently to get her brother to arm wrestle, since they roughhoused often as kids.

"I was holding his hand down and I was like, *Come on, let's arm wrestle* - and he pushed my hand down. So I pushed back and I was like, *Are you going to let me win?* And he pushed my hand back down. That's Eddie - he's very competitive."

Every day Felicia held family snapshots in front of Eddie, and his eyes seemed to fix on them. She pointed out friends and relatives, describing the circumstances behind each image. The family showed Eddie cards and letters from the hundreds sent by friends and strangers. Packages of fruit, flowers and food arrived daily from the Ryans' relatives, co-workers and church members, and the family heard their hometown was planning to dedicate that weekend's Memorial Day parade to Corporal Edward Joseph Ryan II.

Marines dropped by regularly, most of whom had never even met Eddie but wanted to show solidarity. Marine commanders have visited, along with retired Marine snipers. They were upbeat and encouraging.

"It's important that these boys see a positive attitude," Angela said of her son and the two dozen other wounded

Marines on the ward. "We need to give them hope. If you're over their bed crying all the time, they'll know you're doing badly. Eddie always told me, 'Don't worry about me, Mom, I'll be OK.'"

In his room, Bryan Trusty was up and walking, preparing to go home. His doctors had anticipated a stay of several months, but he was being discharged after a few weeks.

"You should have seen this boy," his father said. "He looked like he'd been shot in the face with a shotgun. He went into cardiac arrest on the plane ride home. He had no brain waves. Now look at him."

Bryan shrugged and smiled. Slender and fair-haired, he speaks with a slight Midwestern twang, and said he remembered little of the plane flight - but every detail of the firefight in which he was wounded. It happened during an insurgent attack on Abu Ghraib prison, when he rushed to a guard tower to help fellow Marines.

A rocket-propelled grenade exploded inside the tower, injuring Bryan and five other Marines. In all, forty-four Americans were wounded during the battle.

"I didn't see the RPG come in, but I saw it explode," Bryan said. "I caught a piece of shrapnel in the frontal lobe, and another piece went back to the brain stem." His right arm was broken as well.

A Navy corpsman named Benjamin Graves dragged him to safety. Graves suffered severe wounds himself and ended up down the hall at Bethesda, where he and Bryan were reunited with another Marine injured in the tower.

"When Graves got discharged his mother told us, 'Trusty,

it's time for you to get up out of bed and get out of here too,'" Trusty's father Steve said.

"Bryan's swift recovery is highly unusual, just an absolute tremendous turnaround," said James R. Dunne, the first surgeon to treat him when he arrived at Bethesda. "There is no single reason, just the luck of the draw, really. It depends on the path of the fragment, the extent of the damage - a lot of factors. But these young guys have very plastic brains, and can overcome some really serious injuries."

Bryan felt strong enough on Mother's Day to take his mother, Deborah Hall, to dinner at a restaurant. The following week he was sent to a Veterans Affairs hospital in Louisville, Kentucky near the family home in Indiana for therapy to help improve his short-term memory. He wants to return to school to study computer science.

Bryan and Eddie joined the Marines in response to the September 11 attacks and were based at Camp Lejeune, but they had never met. Even so, before Bryan left the hospital, he sought out Christopher Ryan.

"I told him to just be patient, Eddie will get better," he said. "I said it may look bad right now, but they need to keep the faith and everything will be OK."

Bryan's mother told Angela, "We've seen amazing things happen here, and so can you."

In Eddie's room, Christopher looked down at his son, who seemed to gaze back. "Look into his face and you'll see he's a good boy who helps people," the father said. "He's a selfless person. He was more concerned about his country, and his fellow Marines, and his family than he was about himself. He knew he was fighting for us here at home. Fighting is what got Eddie in here, and fighting is what's going to get him out."

"That - and prayer," Angela said.

As they monitored Eddie's vital signs, the nurses fussed over him and spoke gently to him. The food tube gurgled and Eddie coughed. His mother and his father held his shoulders to comfort him, and another quiet afternoon passed as the Ryan family awaited a miracle.

And then, weeks later, they got it - or at least an early installment. Eddie was stable enough to be transferred to a VA hospital in Richmond, Virginia for rehabilitation therapy. He is now able to move his hands and hold up two fingers on command. He no longer needs a feeding tube, and recognizes friends and family members. He is alert and responsive. He has smiled for the first time since he was wounded. And then, Christopher said, his son managed to speak his first word - "Mom."

**A version of this story first appeared in the *Los Angeles Times***

# FORTUNE FAVORS
# *The Strong*

Steve Dundas

*The author spent nearly eighteen years in the Army and became a Major in the reserves before giving it up to join the Navy as a Lieutenant in the Chaplain Corps.*

I had served as a Platoon Leader, Company XO and CO, Battalion and Brigade staff officer and chaplain in the Army, and I now have had the privilege of serving with the 2nd Marine Division during the past two years. Due to operational needs I have been in three different battalions during my tour, including 3rd Battalion, 8th Marines. They are the Battalion that in 1995 rescued Scott O'Grady, and in 1996 kept the mission in Haiti from becoming a complete fiasco as they were willing to use force against the thugs - unlike the Army, who were literally pushed around.

In 1998 they were the first U.S. troops into Kosovo, and many do not hesitate to say the truth about the nature of the Albanian thugs, having had firefights with them as well as the Serbs. Just before I came aboard they had the politically incorrect mission of clearing Vieques of protestors, and since I have joined them we have completed the less glamorous mission of being forward deployed to Okinawa, Japan and Korea, including some time at Warrior Base on the DMZ.

While in Korea we worked extensively with the ROK 1st

Marine Division. Their officers, observing a bi-lateral Battalion on Battalion FEX were impressed with the way a Marine Corporal could take a squad on an independent operation at night, and how other junior NCOs and Marines took initiative in ways which they had a hard time imagining. The ROK Marines are a tough bunch, and our boys impressed them. This is a proud battalion, led by motivated and competent leaders. As a Chaplain, it was refreshing for me to see Lance Corporals and Corporals serving as Rifle Team Leaders and Squad Leaders, showing leadership in a lot of outstanding ways which I seldom saw in the Army.

We Chaplains often see Marines, Sailors, Soldiers or Airmen as a last resort, after the chain of command has given up on them - at least that was the case during most of my time in the Army. These young leaders, on the other hand, often sought my help early on to get these kids back on track before they became discipline problems and ended up going to the brig or getting kicked out. It was nice to be able to preach (even in counseling) the virtues of taking responsibility for one's actions: being faithful, courageous, and a man of honor and integrity.

My tour with 3/8 was one of the best I have ever had. I seldom had to visit the brig, and didn't have to bury any of my Marines. From the Battalion Commander and Sergeant Major on down, I saw great leadership. Thank God for the Marines, and God Bless the 3rd Battalion, 8th Marines. Our motto is "fortuna favet fortibus," or "Fortune Favors the Strong." It is true.

# GENERAL KRULAK
## *On Character*

*The first time I heard former Commandant Charles Krulak tell this story was when he was the guest speaker at my SNCO Academy graduation, when he was a Brigadier General and the CG of 2nd FSSG. Needless to say, his words left a lasting impression.*

"There is a topic that I would like to talk about today - one that is critical to each of us, our nation, and our world as we move toward the 21st Century. A topic that rarely gets talked about in forums such as this, which makes it all the more important to discuss. It serves as the foundation for all that we are, all that we do, and all that we will be. I will talk about the importance of character.

I can tell you from personal experience that combat is the most traumatic human event. It strips away an individual's veneer, exposing their true character. If a character flaw exists, it will appear in combat - guaranteed.

This morning, I will tell the story of an American whose true character was tested and exposed in the crucible of war. I will then draw some conclusions that are applicable to how the rest of us should live our lives... lives where combat will hopefully never play a role. He was a nineteen-year-old Marine - about the same age as most of you in the audience this morning. His name was Lance Corporal Grable. He was

57

a man of courage... a man of character... and this is his story. In Vietnam it was 0600 on the third of June, 1966. I was in command of "Golf" Company, Second Battalion, First Marine Regiment. I was a First Lieutenant at the time, and had been given this command because the previous commander had been killed about one week earlier. My Company had been given a simple mission that began with a helicopter assault. We would land in a series of dried-up rice paddies about six football fields in length, and three football fields in width. These paddies were surrounded by jungle-covered mountains, with a dry stream bed running along one side. We were supposed to land, put on our packs, and do what all Marines do - find the largest mountain, and climb to the top. There we would put ourselves in a defensive perimeter to act as the blocking force for an offensive sweep conducted by two battalions.

The helicopters landed, unloaded my company of Marines, and had just started to leave when the world collapsed. Automatic weapons, mortar fire, artillery - it was hell on earth. Fortunately, a good portion of my Company had managed to move into the dry stream bed where they were protected from most of the fire. However, one platoon had landed too far west to move immediately to the cover of the stream bed. As they tried to move in that direction, the fires on them became so heavy they had no alternative but to hit the deck. One particular squad found itself directly in the line of fire of a North Vietnamese 12.7mm heavy machine gun. In a matter of seconds, two Marines were killed and three were seriously wounded.

As I watched what was happening from my position in the stream bed, I knew it was just a matter of time before that machine gun would systematically "take out" that whole platoon - squad by squad. If I didn't act immediately, they would be lost in just a matter of minutes. I made a call to the commander of the first platoon which had made its way into the stream bed, directing him to move up the stream bed so he could attack across the flank of the gun position, rather than having to assault it directly from the front. At the same time, I directed another platoon to provide suppressive fire which might diminish the volume of fire coming from the machine gun position. All of this was happening in the midst of smoke, multiple explosions, heavy small arms fire, and people yelling to be heard over the din of battle.

Suddenly my radio operator grabbed me by the sleeve and pointed toward the middle of the rice paddy where a black Marine - a Lance Corporal by the name of Grable - had gotten to his feet, placed his M-14 rifle on his hip, and charged the machine gun while firing as fast as he could possibly fire. He ran about forty meters directly toward the machine gun and then cut to the side, much like a running back might do during a football game. Sure enough the machine gun, which had been delivering heavy fire on his squad, picked up off of the squad and began firing at Grable. Seeing the fire shift away from them, the squad moved immediately to the cover of a small rice paddy dike - thick ground about a foot high separating each paddy from the other. Both they and the other two squads were able to drag their casualties and gear to the position of safety behind this dike.

Grable didn't look back. He didn't see what happened. He kept on fighting. He dodged back and forth across these paddies, firing continuously. He would run out of ammunition, reload on the run, and continue forward - dodging back and forth as he ran. BAM! Suddenly he was picked up like a dishrag and thrown backward after being hit by at least one round.

The rest of the platoon charged. My radio operator grabbed me again, but said nothing. He just pointed to the middle of the rice paddy. Lance Corporal Grable had gotten to his feet. As he stood, he didn't put the rifle to his hip. Instead he locked the weapon into his shoulder and took steady aim - good sight picture, good sight alignment - and walked straight down the line of fire into that machine gun.

About four minutes later, my command group and the rest of the unit finally arrived at the now-silent machine gun position. There were nine dead enemy soldiers around the gun, and Lance Corporal Grable was draped over the gun itself. As only Marines can do, these battle-hardened young men tenderly picked up Grable and laid him on the ground. When they opened his "flak jacket" he had five massive wounds from that machine gun. FIVE.

About seven months later, I traveled back to Headquarters Marine Corps in Washington and watched the Commandant of the Marine Corps present Lance Corporal Grable's widow with the nation's second highest decoration for valor - the Navy Cross. In this woman's arms was the baby boy Grable had only seen in a Polaroid picture.

Grable had displayed great physical courage, but somewhere in his character was another kind of courage as well - moral courage. The courage to do the right thing. When he had the chance to do something else, he chose to do the right thing. His squad was in mortal danger. He had a choice to make, and he did what was right, at the cost of his life. Let me remind you, this was 1966. Grable was a black Marine from Tennessee, who couldn't even buy a hamburger at the McDonald's in his hometown.

Think about Grable... moral courage... personal courage... and character. So, what of *your* character? Who are *you?* No, not the way you look in the mirror or in photographs... but who are you *really?* What do you stand for? What is the essence of your character? Where is your moral compass pointing? Which course do you follow?

Every day we have to make decisions. It is through this decision-making process that we show those around us the quality of our character. The majority of the decisions we have to make are "no brainers." Deciding what we are going to have for breakfast is not going to test your character. Judgment maybe, but not character. The true test of character comes when the stakes are high, when the chips are down, when your gut starts to turn, when the sweat starts to form on your brow, when you know the decision you are about to make may not be popular - but it must be made. That's when your true character is exposed.

The associations you keep, the peers you choose, the mentors you seek, the organizations you affiliate with - all help to define your character. But in the end you will be judged as an individual, not as part of a group.

Success in combat, and in life, has always demanded a depth of character. Those who can reach deep within themselves and draw upon an inner strength, fortified by strong values, always carry the day against those of lesser character. Moral cowards never win in war. Moral cowards never win in life. They might believe they are winning a few battles here and there, but their victories are never sweet, they never stand the test of time, and they never serve to inspire others. In fact, each and every one of a moral coward's supposed "victories" ultimately leads them to failure.

Those who have the courage to face up to ethical challenges in their daily lives will find that same courage can be drawn upon in times of great stress, in times of great controversy, and in times of the never ending battle between good and evil.

All around our society you see immoral behavior - lying, cheating, stealing, drug and alcohol abuse, prejudice, and a lack of respect for human dignity and the law. In the not too distant future, each of you is going to be confronted with situations where you will have to deal straight-up with issues such as these. The question is, what will you do when you are? What action will you take? You will know what to do? The challenge is - will you do what you know is right? It takes moral courage to hold your ideals above yourself. It is the defining aspect. When the test of your character and moral courage comes - regardless of the noise and confusion around you - there will be a moment of inner silence in which you must decide what to do. Your character will be defined by your decision, and it is yours and yours alone to make. I am confident you will each make the right one."

Three years later, in 1969, another incident occurred which shaped the future Commandant's outlook. He was married, had one son and was, in his words, "pushing recruits in San Diego" when he received a letter with a set of orders to return to Vietnam a second time. After his wife dropped him off at the hangar, Krulak struck up a conversation with a young Marine who lived only a few blocks from his house. PFC Cameron was an only child who had never been anywhere, and was petrified.

He said he was a radio operator," Krulak said, "and I told him not to worry, because he'd be in the rear with the gear."

Krulak saw Cameron several times on the trip to Da Nang, and upon his arrival he was met by his unit first sergeant, who introduced him to every Marine in the company. On Day Four, the unit got the "BNGs."

"They were the brand-new guys," he said. "And there was PFC Cameron. If looks could kill! The only words out of his mouth were 'Sir, so this is in the rear with the gear?'" Krulak decided that God had meant for the two men to have a relationship, so he made Cameron his RTO.

"This relationship was bonded by fire, bonded by stress for the next nine months," he said. "And at the end of those nine months, we were given a simple mission."

On that simple mission, a bullet whizzed by Krulak "and right into that precious kid." He dropped to his knees, applied a bandage to the huge wound and cradled Cameron in his arms until something hit him hard on the side of the head.

"The first sergeant had beaned me," he said, "and asked me what the hell I was doing."

The rifle company had continued to fight its way up the ridgeline and Krulak, the commander, was out of the fight. The first sergeant took over with Cameron, and Krulak went up the hill. When they met again at the command post the first sergeant apologized, and with tears running down his cheeks told him Cameron didn't make it.

The first sergeant, a huge man and a veteran of Korea, was only six months from retirement. "He put it all on the line by assaulting and battering his company commander," Krulak said. "But he saw that his commander had lost it, and he had to get my attention. He had the moral courage to take action when he saw something was wrong."

Then there was Krulak's friend, Captain Tom Drowdy, who took a squad leader into a deadly village to find a missing Marine and incurred the wrath of his battalion commander. The commander went "ballistic," and told Drowdy he would be relieved if he did something like that again. Drowdy promised he would not, then turned to walk away.

"He took a few steps, then turned," Krulak said, "and said he would do it again if he had to. The battalion commander relieved him."

Drowdy started to walk to his executive officer to turn over the company when the assistant division commander arrived by helicopter. He had listened to the fight on the radio and told Drowdy he was proud of him and his Marines.

"And he told the colonel that with commanders like Drowdy, he probably had the best battalion in the division." Drowdy kept his job, and went on to become the assistant division commander of the 1st Marine Division in Operation Desert Storm.

"What you need to know," Krulak said, "is the Marine Corps was his life. He did three tours in Vietnam, two Silver Stars and four Purple Hearts. He bled scarlet and gold, but he was willing to give it all up in three steps... but Drowdy got back the only thing he owned in those three steps - his integrity. Only you can give it away. No one can wrestle it away from you. Character is a choice, and that's the big difference. You choose and build it by making hard decisions."

"If you make enough of the tough decisions that made your palms clammy and your stomach churn, you'll be able to find the inner strength to be successful and victorious when the time comes. Leaders are in the inspiration business. They inspire ordinary people to do extraordinary things. That's what your nation is asking you to do."

**Adapted from remarks by the 31st Commandant of the Marine Corps, General Charles C. Krulak, for the Pepperdine University Convocation Series on 14 October 1998 and the Combat Leader Speaker Program at Fort Benning, Georgia on June 22, 2004.**

# K2

*The first time I saw then-Captain Rodney Richardson was when I arrived at Camp Talega, at the northernmost end of Camp Pendleton, to check into 1$^{st}$ Recon Battalion in 1983. He was the CO of Company B, and looked to be as hard as petrified woodpecker lips - and as I came to learn in the next couple of years, that was an accurate assessment. The bottom line is, there is no way a terrorist - or ten of them for that matter - could have taken him in a fair fight.*

Rod "K2" Richardson died near Baghdad, Iraq as a result of injuries received from an improvised explosive device, while serving his country as the civilian manager of a private security company.

He was born on October 3, 1952 in Lamar, Colorado, was reared and educated in Kansas and Oklahoma, and graduated from Boise City High School in Oklahoma in 1970. Following high school graduation he enlisted in the Marine Corps and served in Vietnam, and after returning attended and graduated from East Central University in Ada, Oklahoma.

Rod re-entered the Corps after graduation and attended Officer Candidate School in Quantico, Virginia where he was commissioned as a Second Lieutenant. He held ever increasing assignments and responsibilities with the Fleet and in the Far East, mixed with shore duty and Marine schools in the United States, and was selected for and served a two-year tour of duty with the British Royal Marines in England.

Throughout his years in the Marine Corps he was an accomplished parachutist and diver, and a leader in training and fighting in the jungles, the desert and with the fleet, including some operations still not known to the public.

He developed a love and ability for teaching mountain climbing and warfare tactics at the Pickle Meadows training site in the Sierra Nevada Mountain Range, and later this climbing ability allowed him to ascend several of the highest peaks in the Himalayas and elsewhere. In fact Rod and his wife Rita built their home near the base in the Sierra Nevada Range and resided there until his retirement from the Marine Corps in 1994.

Following his retirement he began work with Rite of Passage, a program designed to assist youths with discipline, education and training, and also completed his Masters of Business Administration degree at Chapman University. Then, shortly after the war began in Afghanistan, he signed on as a contract employee providing VIP security services. He said, "I was trained for it, and I want to help!" As the war spread to Iraq he continued his work in Iraq. Besides brief periods of returning to the United States to spend time at home with his wife he served in the midst of unconventional war zone from the onset.

Rod strongly believed in service, discipline, physical training and developing the abilities of those he commanded. He cared for his men, but did not expect them to do anything he could not do. He was known for always being out front as a leader.

There are two circles that form when someone dies. The inner circle consists of the family and friends of that person - those who knew them best and loved them dearly. The outer circle consists of the rest of us, and it is from that perspective that I mourn the death of Rod Richardson. I remember when he was simply known as "one of the Richardson kids" back when he played sports, back when he dazzled the girls, back when life was simple and "dragging Main" was the thing to do. And long before he became known as Lieutenant Colonel Rod Richardson, USMC, or by his nickname " K2."

I remember Rod working alongside his siblings - Deb, Pam, Mike and Sandy - at their family restaurant, the "Cozy Café." The memory is still so clear. Their dad Gene is standing in the kitchen, frying the best hamburgers in town. Their mom Reta is standing behind the counter, a pencil caught in her red hair. The kids are waiting on customers and bustling back and forth, bussing table and handing out menus. I imagine that it was there, working beside his family, that Rod learned how to be a team player and a leader - two abilities that would serve him well in his lifetime.

Throughout the following years I would hear snippets about Rod and the life he was living far from our city limits. I knew he had married a girl named Rita, he had joined the military, traveled the world and was one of the "good guys" watching out for the rest of us. But I never saw him again. Not until the wee hours of Monday morning, when I found myself looking at a picture of him on the Internet, entitled

"Hilla-Najaf." I don't know when the picture was taken, but his face was so familiar, his hair still brown though his temples were graying, the handsomeness of his youth still evident. As I looked at the picture and as I read what his comrades had to say about him I wished I had known him, too. Rod was so respected and loved by the men he commanded and the Marines he fought beside. He was honest and forthright, he instilled confidence in those around him, and he was an example of life lived with passion.

From my place in the outer circle, I watched the inner circle gather together over the past week. They came from many places to say "goodbye" to Rod. Friday evening I drove by the mortuary, wanting to stop and pay my respects, but not doing so because there were so many people already there. Out in the parking lot there were clusters of men talking and laughing, celebrating their friend's life even as they were mourning his death. I hoped that some of them were looking at the western sky, seeing the world that Rod once lived in. Night had already fallen, but there remained a ribbon of orange on the horizon. And up in the darkened sky there was but one star visible, one star shining down on that parking lot.

And then it was Saturday afternoon, and with my family I was standing on the side of a street that Rod had traveled so many times in his youth. Each of us were holding an American flag, the symbol of Rod's passion and sacrifice, as the hearse, his family and his friends passed by on their way to the cemetery. I wondered then if Rod would have been surprised at the number of people who were honoring him that day.

I continued to feel the need to step beyond the outer circle, and so I went to the cemetery Sunday afternoon. It was quiet out there. It was peaceful, it was windswept and the sky was overcast. I knelt down beside Rod's grave, beside the many arrangements of flowers that covered the mound of dirt. I thought of him, his mom and dad, his siblings, his wife, his friends. I could only imagine the adventures he must have had on his journeys to distant countries. And I wondered about the people whose lives he had touched and whose lives he had saved. For a moment, I thought I heard "Taps" being played as it was at his funeral, its echo coming from somewhere far away, perhaps from Iraq or Afghanistan. And I found that the only words I could speak out loud at that moment were "Thank you, Rod." As I stood up, I plucked a solitary yellow rose from the many laying there, and as I held it gently in my hands I thought how Rod was like that rose - just one of the many, many men and women who have become our fallen heroes. And as I looked towards town - the place that Rod once called "home"- I remembered the words "What we do for ourselves dies with us. What we do for others and the world remains, and is immortal." You will never be forgotten, Rod.

**This story has been adapted from an article in the Boise City, Oklahoma news which was written by one of Lieutenant Colonel Richardson's high school classmates.**

# G. I. JOE

**Bill Hoover**

*If you ever wondered what it was like to be a Marine during the battle on Saipan, this will make it clear. In one chilling paragraph, Bill Hoover puts you there.*

It was about D-Day-plus-5 on Saipan, and I was watching for Japanese soldiers along the beach who were sneaking in to give directions for artillery fire. I was sitting in the turret of an armored Amphib, and around 2300 saw the faint outline of what appeared to be a Marine walking along the beach. There was no moon, and the only light was from a fire burning about a hundred yards away.

Japanese soldiers had a habit of wearing U.S. Marine helmets and carrying an M1 when they could find them, and in the dark they were hard to identify by a silhouette. When this guy got to within about fifty feet of me, I asked the usual "Who goes there?"

The answer came back in perfect English. "IT'S ME, GI JOE. A MARINE LIKE YOU." So I aimed about eight inches below the outline of his helmet and shot him, and the next morning there was one dead Japanese officer laying in the sand. I had never heard of a Marine calling himself "GI Joe." I don't think I ever will.

# GOD AND GUNS

**Scott Peterson**

The first few pages of Marine Corporal Tim Milholin's small zip-up Bible are stuck together - drenched "too many times" from the sweat of battle, he explains. It lives under his armored vest in his chest pocket, with an inscribed metal plate which says, "The Lord is my strength and my shield." Corporal Milholin carries a Bible, photo of his wife and an inscribed religious medal into battle.

Milholin is girded for war in Fallujah with both book and sword. He is as well-versed in the King James text as he is in the killing potential of hollow-tipped bullets or the amount of C4 plastic explosive and TNT needed to blow through an Iraqi door. To him, they are all essential tools of his warfighting trade, as important as the photo of his wife, Brianne, that's tucked inside his helmet.

"I pray earnestly every day, and believe that God puts his angels out before us, to protect us," says the Marine, who fires up his camp stove daily before daybreak to brew coffee for the unit during the violent days of Operation Dawn in Fallujah. Since the dark night when they rolled into the dense urban environment of this now-empty city of 300,000, U.S. forces have been in their toughest fight since the Vietnam War. As they search for their enemy, breaking through one closed door after another, the Raider platoon - the Death Dealers, as they dub themselves - are on the front lines in a city hammered to rubble.

They're a microcosm of the modern military, a disparate handful of young men drawn from the melting pot of America - but they share obsessions... with guns, God, guitars, girls, wives, and fiancées. Most took part in the 2003 invasion of Iraq, and the common experience of combat has deepened a bond of brotherhood - a tie upon which their lives depend every day on the terror war's most dangerous battlefield... Fallujah. In this crucible, they have seen death and delivered it, and grown mature beyond their years amid unrelenting rigors and danger.

Every day, sometimes twice or more in a twenty-four-hour period, the scouts gather for final orders. The moment of deepest contemplation comes before each attack, often early in the morning, as on the group's seventh day in Fallujah. In near silence and darkness they clean weapons once more, pack rifle magazines with bullets, and load gear belts with explosives.

Not all are religious, but a few scouts - like Corporal Milholin - keep small Bibles in their chest pockets, close to pounding hearts. Many use a black permanent marker to ink their hands or gloves with their blood type and "kill" numbers - information that will enable news of casualties to be passed immediately over the radio. It's a habit that's taken on greater significance in the course of a month of battle that has killed or wounded more than twenty percent of Charlie Company, 1st Light Armored Reconnaissance (LAR) battalion.

Not all are impressed. "I don't write any of that crap on

me," says Lance Corporal Matt McClellan (X58, B+), a tattooed serial rule breaker. "It's bad luck."

It was just August when the company commanders created the concept of Raider One - a single vehicle that can deposit up to ten scouts on the ground within seconds to fight in conjunction with Light Armored Vehicles known as "war pigs." The setup provides new flexibility during hand-to-hand combat and has proven so effective that Raider is assigned constant missions in Fallujah.

It's been during these operations that the brutal emotions of battle, of tragedy and triumph and coping, mix with Washington's calculation that Fallujah - which was a hub for hostage-taking, rebel weaponry, and car bombs - had to be destroyed, in order to be saved.

"This is urban combat to a 'T,' with 360 degrees of danger," says Sergeant Kevin Boyd, the young-faced chief scout who forged Raider's clockwork skills of house clearing by daily practice on the ship to the Middle East, storming stairwells and clearing catwalks on upper decks.

"You've always got to be looking in every house - behind every couch there could be a guy hiding," says Boyd, an Eagle Scout who wore his first camouflage at age three and owns more than twenty guns. Boyd graduated from high school on a Friday, celebrated on Saturday, and left for the Marines on Monday. He says Fallujah is "ten times" as dangerous as the Iraq invasion, during which LAR lost only one Marine - who stepped on an artillery shell.

"It's a lot faster combat, a lot more deliberate. Grenade, grenade, rocket-boom! You're in," says Boyd. For luck, he keeps an Ace of Spades in his helmet.

"I love the adrenaline of it, the fast pace," Boyd adds. "I'm breathing in plaster and composition B from the grenade, choking on it - spitting out black stuff as I'm clearing the room out. It's great!"

"There is nothing more personal than someone trying to kill you, and you trying to kill him," says Captain Gil Juarez, the LAR company commander. "Not marriage. Not parenting. Nothing is more personal than having to toe the line, when it's either you or him."

During pauses between operations the men set up camp, living cramped in occupied Iraqi houses. It's at such times that the Marines try to digest the unpredictable moments of Fallujah, with feisty debates that erupt about everything from too-young girlfriends to the utility of God.

"I'm sure they see it in every war - so many people become religious out here," says Milholin, sitting on a floor mattress covered with red satin. The windows of the house are gone - smashed out by the Marines so that glass will not fly when mortars land nearby.

Milholin and two others are known as the "Three Wise Men." "I put so much faith in God, I don't know how people do it without being religious."

Corporal McClellan knows how - and often takes issue with the Wise Men's certitude. "I have confidence. Ever since I was a little kid, I knew I was not going to die, so I don't need [religion] to lift me up," says the machine-gunner. McClellan racked up twenty-six counts of grand-theft auto while still a juvenile. He had six ear studs on one side, seven on the other, and a tongue stud - which once got stuck in his lip ring. Joining the Marines has tempered such behavior, but it hasn't erased McClellan's independent streak.

If anything, McClellan says he blames God for what goes wrong - a key reason being the fate of his friend Lance Corporal Kyle Burns, of Red platoon, who was going to be the best man at McClellan's wedding until he was killed in an ambush. McClellan took the death hard. He stills clings to a photo of himself and the square-jawed Marine in a cowboy hat from Laramie, Wyoming, sitting together smoking a Middle Eastern water pipe.

After the ambush, McClellan was put in charge of guarding an Iraqi who had surrendered with a white flag. The Marine made no secret of his distaste for the man. A second Marine was added to guard duty to keep an eye on both of them. "If anybody left me alone with him, I would be in the brig right now," McClellan says later.

Losing his friend changed McClellan's sense of mission. "Before [Burns] was killed, I thought we were here to kill the bad ones and save the good ones," says McClellan, a wry wit who sometimes jokingly refers to himself as "trigger-happy Mick." "Now I think, 'Is he the one who shot Kyle?' It's a revenge thing. Every time I see an Iraqi, I could be face-to-face with the guy who killed my best friend."

Perhaps as a result, McClellan expects this conflict to bring him closer to his father, a former Marine who survived three tours of Vietnam unscathed and fought in the urban battle at Hue City in 1968.

"He used to talk about how, ten seconds from now, you

don't know if you'll be alive," McClellan recalls, swinging a pair of dog tags on a chain. "His buddy right next to him was shot in the face and died. Now I know how he felt."

Burns' death has become a point of debate within the unit. The Marine shared McClellan's animosity toward religion until just days before the ambush, when he "gave his life to Christ" at a church service, according to some who were there.

"God has a perfect plan," says Corporal Christopher DeBlanc, a team leader and one of the Wise Men. He keeps a red leather Bible in his rucksack, part of a pile of personal gear deposited upstairs in the Iraqi house. "For example, Red platoon got hit by a mine, and after that they had a church service. They accepted God. Kyle accepted it. Kyle is in heaven now."

*"That* was God's gift to Kyle?" asks McClellan incredulously. "Great. You accept God, and the next day you get killed. That's some advertisement. You are done at twenty years and three months, unmarried."

That reaction doesn't surprise Corporal DeBlanc, a tall, reliable Marine. His path to the military, and to his overarching faith, has been circuitous. From the age of twelve he worshiped the guitar and played in rock bands, practicing for hours after school, and often falling asleep with the instrument in his hands - but his rock-star lifestyle didn't take him far. "During my teenage years, I hit the bottom of the barrel," DeBlanc says. "When I joined the Marine Corps, I was done living that way."

His change of heart was sparked by the burial, with full

military honors, of his grandfather - a World War II veteran and a man he wanted to emulate.

"I didn't cry at first, but when the honor guard got up there for the twenty-one-gun salute... as soon as that first round cracked, it was Niagara Falls," DeBlanc recalls. "I didn't like how my life was going, so I gave it to the military."

The Marines instill a new set of values and "force you to grow up," says DeBlanc. For him, that included a growing framework of faith that he applies in Fallujah. "The big thing is the spiritual battle going on in our lives - the fight we're fighting is good against evil," he says. He knows the Americans are not the only ones to call on divine power. On the wall of one house, written with yellow paint in Arabic, it says, "God help [Iraq's] mujahideen."

DeBlanc easily reconciles war with the biblical commandment against killing. "Doesn't the Bible say, 'There is a time to pick up the sword, a time for peace, and a time for war?'" he asks. "I can pull the trigger here and have a clear conscience."

To a degree, that goes for Lance Corporal Jason Bell, the original Wise Man, who tries to balance the battle with the message of his Bible, which he keeps on him in a plastic meal sleeve that also holds a stun grenade and an extra rifle magazine.

"I always prayed, before we came here, that Iraq's innocent civilians wouldn't look badly on us," says Corporal Bell. He wistfully recalls breaking out in laughter with a young Iraqi after several failed attempts to communicate - an uncommon moment of levity between Americans and Iraqis.

Bell's faith was tested during a pre-dawn raid, when small fragments of a U.S. grenade ricocheted and embedded in his cheek, effectively shielding this correspondent from the blast.

"I thought it was a blessing in a weird way - the wound wasn't that bad," recalls Bell, who wants to go to Bible college and preach. "It's kind of crazy. God told David he couldn't build a temple, because he had blood on his hands. Though we've been in contact, I don't know that I've killed anybody. I've never hesitated, but it seems whenever I've gone out, there was nothing out there."

DeBlanc also plans to go Bible college, and has dreamed of himself as an elderly man at a pulpit, his wife with three children (as yet unborn) in the front pew. That's a welcome change from the nightmares he had for a year after returning home from Iraq in 2003. "I would wake up, looking for my rifle," says DeBlanc. "I dreamed I was in a fire hole and being overrun, and couldn't find my weapon."

"Doc" Nick Navarrette, the U.S. Navy medical corpsman from Omaha who serves as Raider's ambulance chief during casualty evacuation, had a nightmare too in Fallujah. "There are fifty Iraqis coming at us, and I had an M-16, and all it shot is dust," says the slightly built corpsman. "I reached for my pistol, and it's only crunching sand."

The corpsman's job requires him to be a noncombatant, limited to using his M-16 rifle - but when the Raider One vehicle was sent to reinforce Red platoon during the November 11 ambush, he got behind a belt-fed machine gun when he saw three insurgents shooting from a third floor. He killed at least one, and stopped fire from the others. Then, when casualties were announced on the radio, Navarrette's real work began. After he gets home from Iraq, he wants a fourth tattoo - a pair of angel wings across his back.

"As soon as I jumped out of the vehicle, the shots started flying by my head. I could feel the wind from the bullets in my face," Navarrette recalls of his thirty-yard dash to the wounded. He got the first casualty back to the vehicle when word came about another one. Navarrette and a gunner, bullets striking in front of their feet, retrieved Burns - who was dead.

There is speculation among Raider platoon that Navarrette may be put up for a Bronze Star with a 'V' for combat valor, but his mother was angry. "She told me, 'Don't do anything crazy,'" he recounts after a call home. "I'm already a hero to her; I don't need to be a hero to anybody else. I tell her, 'It's my job. These are my friends.'"

Days later, wrapped in McClellan's thick blanket to ward off a morning chill in the occupied house, Navarrette elaborated while eating potato chips and French onion dip from a care package. During the call to his mother, he also learned that a close friend, Shane, was killed the same night. The news also reached Shane's wife, April, just hours after she gave birth to the couple's first child. The emotional ride gave the corpsman pause, and he pulled out the video camera he used to film part of the ambush.

The rattle of bullets and pounding high-caliber rounds dominate the soundtrack. Then the footage goes calm, and

Navarrette is speaking from the back of the Raider vehicle after delivering the casualties to the combat hospital.

"Yeah, well, we lost two guys. I'm here by myself," he records, clearly shaken. "I got blood all over my hands. I got blood all over my pants, and my flak vest. It was not a good day. I never want to go through a day like that again."

"But we're going right back out there. No breaks," Navarrette says, his voice threadbare. "I'll turn you back on, when the bullets start flying again. All right - peace."

**This story originally appeared in *The Christian Science Monitor***

# SUCH A DEAL
## *I'll Take Two Corps Full*

Colonel David H. Hackworth

"Battlefields seldom change," I thought as I walked the perimeter of the Marine Basic Course and observed the deep foxholes, outposts, barbed wire, fields of fire, wet, alert young warriors, ankle-deep mud and always, the smell of gunpowder.

Here John Glenn, Chuck Robb, and my cousin, former Navy Secretary Jim Webb, learned the basics of leading men and winning in battle, as did California's Governor Pete Wilson and tens of thousands of other patriots who joined up to serve with America's finest. None of the Vietnam-era presidential hopefuls passed through this crucible. They all dodged the draft to serve a higher priority - themselves.

At Quantico, Marines learn not just to kill, but to lead, to think and to absorb standards that stick with them for life. Character is forged in an environment where perfection is not good enough, where duty, honor and country are forever grafted onto their belief systems. That's why so many Marines lead the way in almost every pursuit in this land.

There's little difference between the current crop of Marines and the "Devil Dogs" I first met as a ten-year-old shoeshine boy in 1940. Then, too, they were sharp, salty and proud - and they liked to keep their mahogany shoes glistening, which was good for business. They were not in the Corps because it was a job. They had joined up because for them it was a near-religion, a compelling call to serve their country.

As I watched the kids who still have that calling dig in, I thought, "Nothing has changed since before Pearl Harbor." The faces are still young, the minds eager, the bodies rock hard and the equipment clean and serviceable, though worn and old... very old.

The big difference between Marines and the Army, Air Force and Navy is the Corps runs on the smell of an old oily rag. They're the poor cousins of the other, richer services. Colonel James T. Conway's total annual budget for putting almost three thousand officers through Basic School is a lean $967,031 per year. The Army's "kiddieland" at Fort Bragg, built to baby-sit serving soldiers' offspring (71% of the family-oriented U.S. Army is married), costs five times as much. A month's per diem (hotel and food) for three hundred USAF fighter jocks in Italy - who are too princely to sleep on cots in tents as Marines do - is about one million a month. The cost for a headquarters in Naples to deal with ex-Yugoslavia is eight million a year and boy, do the staff weenies there live high on the hog.

The Corps gets only six percent of the defense budget. This pays for twelve percent of the active forces, twenty-three percent of the active divisions, thirteen percent of the fighter/attack aircraft and fourteen percent of the total reserve force.

It doesn't take a whiz kid to figure out this is one hell of a lot of bang for the defense buck. Marines don't waste defense dollars. They're into lean meat, not blubber. Quality of life to leathernecks isn't pampering and frills, but a resupply of ammo on the high ground.

Defense Secretary William J. Perry knows his budget will be halved by the year 2000, leaving us with a broken defense machine. The Pentagon has got to trim now to be able to fight later.

Perry should find out how the Corps can do so much with so little and ask, 'Why do Marine pilots sleep in tents next to their planes, while Air Force pilots live downtown in plush hotels? Why does the Army have two hundred major generals for only ten divisions? Why do Marine sergeants serve as navigators aboard Marine C-130 aircraft, while majors do the same job in the Air Force? Why does the Corps have one officer to every nine Marines, when the Air Force ratio is one to four, the Army one to five and the Navy one to six? Why does the Pentagon have more people now for a force of only 1.6 million than it had in 1945, when the force was thirteen million?

The Corps is one hell of a defense bargain. Pound for pound, in these days when cost-effectiveness is so critical, the Corps provides by far the best value at the best price.

**Used with permission of Colonel David H. Hackworth, USAR. Originally published March 14, 1995.**

# CAPTAIN CHINA

*I had the privilege of meeting "Barney" Schenn through my good friend Gunny J. C. Allen, with whom he had served on the drill field. The two of them developed a lifelong friendship, and when J. C. and I took a trip to California for a Force Recon reunion we enjoyed the hospitality of Colonel Schenn and his lovely wife Mary for an evening - and it was then I first heard about "the saga of the name change."*

Colonel Byron "Barney" Schenn's grandfather, John Trzeczcinski, left Kuczbork, Poland - which was part of Russia at the time - to avoid conscription in the Czar's army. He emigrated to the U.S. around 1893 and settled in Baltimore, Maryland where he met and married another immigrant from Skalsk, Poland named Maria Nowakowski. Being strangers in a new land, they settled in an enclave amongst "their own," which was the custom at the time. Often the Poles would gather at an establishment at Fells Point and marvel at how lucky they were to be in Baltimore eating steamed crabs and drinking beer instead of living in Poland raising sugar beets for making vodka or serving in the Russian Army.

By the time Barney's father was born in 1896 part of the last name had been dropped and the family name recorded on his church birth record was Czcinski, but after 1900 it was listed as Trzcinski - something Barney calls "the Polish way." So it remained until one summer evening in 1910, or perhaps 1912, at the *Admiral's Cup* bar. A conversation started among the patrons about the way immigrants groups were labeled by Americans. The Italians were "WOPS" (an abbreviation for Without Papers), the Germans "Krauts," the Irish "Micks" because their last names were all "Mc-something," and the Poles were all called "Ski" - and they were tired of being called Ski all the time!

A member of the group offered, "We are in a new land, with new customs, so maybe we should shed our Old World identity." Barney's Grandfather John stood up with his glass on National Boh - or maybe it was Gunther's - and stated, "I am going to change our last name for our children." The gathering of Poles asked what the new name would be, and he responded, "Well, as a start I'll cut the 'ski' off the end, and since Americans can't seem to pronounce the 'Trz,' which is mostly silent anyway, I'll cut that off the front - and that leaves 'Cin.' Now, how that would be spelled in American?" Someone said, "I believe it would be C-h-e-n" (although Cheen would have been closer to an accurate Polish pronunciation), and Grandfather John proclaimed, "Sounds good to me... the new name is Chen!"

He was pleased, and so the name was adopted for *some* of the family - although no legal evidence for this change has been uncovered. Grandfather John, however, did not change his name, and Trzcinski is on his tombstone. Maria did change hers, and is buried beside John with a common headstone. Uncle Steve became "Shen," as he lived in Hawaii where there were a large number of "Chens," While

Uncle John kept Trzcinski, and so on. It is easy to see how there is a real problem for anyone interested in tracing the genealogy of this line!

As Barney and his siblings grew up they thought, "If our grandparents were Polish, how did we end up with the name of Chen - which is Chinese?" Others wondered too, but did not ask. Barney's sisters escaped the name by getting married, but was destined to go through a good part of his life with a Chinese last name. How appropriate, then, for him to serve as a Marine in China, Japan and Thailand!

As an eighteen-year-old Private stationed in North China, he did not recall any reaction to his name, but does remember Charley, the company barber (and no, his last name was not Chan, or for that matter Chen!) trying to figure it out. Later, while attending Officers' Candidate School at Quantico, Barney was tagged with the nickname "Captain China" - which has stuck to this day.

In a clothing store in Hong Kong he was asked his name and said, "Chen, of course," and the Chinese man replied by saying, "That's my cousin's name!" Barney said, "It's mine too, and here is my USMC ID card to prove it." He then explained that he probably had bone grafts in his legs to make him taller, and eye surgery to get that Western look. The Chinese examined Barney closely, and while he detected some skepticism he did get free scotch and cigarettes and four shirts and three pairs of shoes for the price or two. It paid to have a Chinese name while shopping in *authentic* Chen territory.

Next he was off to Tokyo to fill a position on the Honor

Guard for the United Nations and Far East Command - duty which required an officer at least six feet tall. "Cancel these orders," the Commanding Officer had said. "This must have been a mistake... Lieutenant Chen couldn't possibly be six feet tall." The Adjutant suggested they wait until he arrived. They did. He stayed.

On Taiwan, while waiting to have dinner with the Commander in Chief of the Chinese Navy, Barney noticed a perplexed Chinese Protocol Officer attempting to finalize the seating arrangements. He could not identify anyone who appeared to be Major Chen so he could show him to his table, until finally the only unseated person remaining in the room was a U.S. Marine officer. Barney solved his dilemma. The officer remained perplexed.

On Okinawa, after returning from Vietnam, Barney was referred to as "my Chinese roomie" in letters home by a "zoomie" quarters mate. Then on to San Diego, where the local Chinese Mandarin Church invited attendance at services. He did not go. Then another letter saying, "We haven't yet seen you on Sunday." They still haven't.

Upon retirement from the Corps Barney settled in Orange County, California, and one day while working with an engineer from San Francisco named Jones Wong he asked, "How did you get the name of Jones, which is a common last name in the U.S.?" to which Mr. Wong countered, "Well I have been meaning to ask *you* about *your* last name."

Chinese mail kept coming - from dentists, markets, school book stores, banks and so on. Barney sent some back with the notation, "Please translate, and return." They did not.

# Been There, Done That... Got the T-Shirt!

Why all the mail? If you check the Orange County phone book, you will find more Chen's than Smith's or Jones'.

Finally, the clincher. A letter came from San Diego State University informing Barney of a study to be conducted on "the effects of environment and culture on the aging of Oriental men." They wanted him to participate, and said they would call for an appointment. He did not respond. That was enough! It was now time for another name change.

Since Barney wanted to keep the pronunciation close to his "Chinese" name for easier identification, and did not want to go the more difficult Polish route, he came up with three different versions and put them on a ballot for his female staff to vote on. Schenn won, and that is what it is today - properly approved by the court, published for record, and noticed to friends. It is the old name with an "S" on the front and another "N" at the end, and is now framed on the wall courtesy of Marine friend Peter Beck.

Now there is no more "Captain China" from the 7th Basic Class, no more "Chinese roomie" from Bill Kull, no more mail from the Chinese Mandarin Church - but what to do with the military citations and plaques with "Chen" inscribed thereon? Barney has since been informed by the former editor of *Leatherneck* magazine that he is still "Captain China" - and a mere name change will not alter that!

So now Barney must decide what will one day go on his grave marker. Trzcinski? Chen? Schenn? Captain China? Or perhaps just, "A MARINE, Semper Fidelis."

Adapted from *The Saga of the Name Changes* by Colonel Byron Schenn USMC (Ret)

# MARINES ARE DIFFERENT

**Paul Akers**

The Marines are indeed different, and thank goodness.

With their airstrip destroyed and shaken by news that reinforcements had turned back, fewer than five hundred Marines defending Wake Island in December of 1941 prepared to meet a Japanese amphibious invasion backed by twelve cruisers and destroyers. Out of such straits legends are born, so when asked by Pearl Harbor headquarters what the besieged Leathernecks needed most, a Marine radioman supposedly replied, "Send more Japanese."

The story is apocryphal, but spiritually consistent with the Corps, which rebounded from the fall of Wake to dislodge a dogged enemy from island after island in some of the bloodiest fighting ever witnessed. On Iwo Jima, the Marines suffered almost six thousand dead in just over two weeks, and casualties were not confined to the lower ranks - nineteen of twenty-four original battalion commanders fell.

Somehow the Marines managed to take Iwo and (with Navy and Army help) and push the Japanese back to their home island without mixed-gender training platoons. If history is any guide, the Corps will be called upon to do similar work again, so Marines cleave to the dictum, "The more you sweat in peace, the less you bleed in war." This means young male Marines will continue to receive grueling training without the dilution or distraction that the presence of female Marines would bring.

The Marines alone resist "coed" training. That adjective is apt because the boosters of such training think it more crucial that America's warriors be re-educated in progressive social fashions than taught, say, how to parry a bayonet thrust. However, the Marines' principled obstinacy gives them a high silhouette for political snipers. Sara Lister, an assistant Army secretary, once fired a few rounds during an academic conference, calling the Corps "extremist" and "dangerous." Dangerous, anyway, to blatherskite bureaucrats - she was forced to resign hours after her vitriol hit the front page.

Although Ms. Lister really did savage the Corps, the proper response was the one given by Marine Commandant Charles Krulak. The "extremist" label, said Krulak, "would dishonor the hundreds of thousands of men whose blood has been shed in the name of freedom. Citizens from all walks of life have donned the Marine Corps uniform and gone to war to defend this nation, never to return. Honor, courage and commitment are not extreme."

Since Krulak accepted Ms. Lister's apology, it was a little stomach-churning to see politicians like House Speaker Newt Gingrich - the hero of a Tulane graduate school during the Vietnam War - demand her head for unpatriotic utterance. Especially since she really was, albeit in a bellicose way, onto something.

There is a vast cultural gulf between the Marine Corps and civiliandom - but then again there is a gulf separating all of those whom Marines are wont to call "non-essential personnel" - soldiers, sailors, airmen, and civilian society. Unfortunately the gulf is narrowing between the other services, and the squishy zeitgeist of the dominant culture.

In the Navy's gender-mixed boot camp, notes the Heritage Foundation's James Anderson, new sailors are issued a "blue card" to cope with stress. Recruits who feel discouraged may turn in their card for soothing words or, if acutely frayed, make collages. Such pop therapy comes atop watered-down training standards.

Kate O'Beirne of *National Review* points out that while one hundred percent of men can discharge the critical on-board task of carrying a one-hundred-forty-pound sailor to safety on a ladder, only about eight percent of women in two-person teams can do so. Might this be a problem on a stricken ship? "No," a Navy officer perkily says, "because we have redefined the task. Carrying a stretcher is now a four-man task." Privacy screens installed to shield female sailors from gawking have even increased the time it takes for seamen to reach battle stations.

In Army boot camp, recruits now run in sneakers because combat boots tend to cause stress fractures in females. (Nikes: Don't go to war without them!) "During the obstacle course," says retired Army Lieutenant Colonel Bob McGinnis of the Family Research Council, "recruits may go around the wall if they can't climb it." Kinder, gentler Army drill sergeants no longer verbally dress down recruits - even as some sergeants have physically undressed a few of the females in their power.

Until last year, says the Air Force chief of training

analysis, physical exercise in that service was so flaccid that some recruits were actually being "*de*conditioned."

When egalitarian ideology and the shoddy value system of the MTV-mall culture invade the armed services, attitudes change dangerously. An *Army Times* survey reveals that just a third of female soldiers and fifty-seven percent of males agree that "the main focus of the Army should be war-fighting." While machismo-brimming Hispanic youths flock to the Marine Corps, the Army's recruitment of young Hispanic men is down fifty percent.

One lesson here is some like it tough, and another is to sign them up, don't put Tommy Hilfiger on a recruitment poster. Or a coed platoon, either. Tell them a story like Wake's, and keep plenty of ink in the well.

**Paul Akers writes editorials for Scripps Howard News Service. This originally appeared on Sunday, November 30, 1997.**

# HISTORY LESSON

*You only need to turn on Fox News or CNN to know this is true. When President Reagan said, "Some people live an entire lifetime and wonder if they have made a difference in the world - but Marines don't have that problem" he was absolutely right. No matter where or when you served, or what you did, you have made a difference in the world - which is something very few of the protected masses can say.*

It seemed from the conversation that this fellow, despite his appearance, had flunked his physical examination and an officer was giving him the bad news. The blonde guy pleaded for an exception which would allow him to become a Marine.

The captain listened quietly, staring straight ahead, thought for a moment, then said, "An exception could be made only if you have some special skills or training that the Corps needs. In that case, we might consider a waiver in order to take advantage of those skills. Tell me, young man, what is your profession?"

"I teach history, Sir."

The captain stared into the blonde man's eyes, and spoke slowly and deliberately. "Son, we don't teach history, we *make* it!"

**Courtesy of *Stories From the Pacific***

# THE AIR FORCE
## *And an Elite Force*

**Kyle R. Fix**

An Air Force Sergeant approached a Marine Sergeant outside an Enlisted Club to voice his disapproval after witnessing the seemingly harsh treatment the NCO had inflicted upon a young Lance Corporal.

It seems the young Marine had over-indulged himself on alcoholic beverages, and had commenced a hands-on demolition of the interior of the building, before being hauled outside via the ear by one relatively large Marine Sergeant. Once outside, the Sergeant had apparently backhanded the Lance Corporal in the back of his head, ordered him lock up his drunken body at the position of attention, and proceeded to verbally reprimand the Devil-Pup in a manner befitting the behavior he had exhibited.

When mission complete with the verbal full frontal assault, the NCO ordered the Marine to return to the barracks and standby for Rocks and Shoals the following morning. Totally appalled by this public display, the Airman (to whom some credit should be given for having the fortitude to do so - although the line between bravery and stupidity is very fine) approached the Marine Sergeant and commented, "Hey, don't you think you were a little too harsh on that young man?"

The Marine very calmly but firmly stated, "First of all,

you'd better execute an about face and commence walking before you spring a leak. Second of all, that's exactly the kinda thing that makes men like me part of an *elite* force and people like you, part of the *Air* Force."

Each party then silently parted, leaving with us yet another "Corps-ism" which has been passed on from generation to generation.

# MORE THAN A FEW
## *Good Men and Women*

**Colonel David H. Hackworth**

Damn. The current crop of Marines is awesome. I recently spent a couple days with our Leathernecks and talked to a fair number of young warriors, most of whom were students at the Marine Corps Amphibious Warfare School at Quantico, Virginia.

America is fortunate to have such bright, dedicated and selfless professionals. Their motivation is to defend America, and if that means living in the mud, being constantly away from home on missions and perhaps dying for the Red, White and Blue, they're ready to pay the price. None are interested in becoming CEOs, owning BMWs or flaunting Gold Cards. They're not members of the greed generation, but rather old-fashioned patriots who want to serve their country, lead its sons and daughters well, and keep them alive.

Since the "Halls of Montezuma" the Marines have been a special branch. Had a Marine sergeant not been such a stickler for details, I would've worn Marine green. I tried to join up in 1944, after a recruiting poster Uncle Sam pointed a finger and said, "I WANT YOU," but the recruiting sergeant said, "Come back next year, kid." I came back in '45, he looked at me hard, but again I got the big rejection to "Try

97

'46." Even as a dumb kid of fifteen, I knew the war would be over before I got big enough to pass the Gunny's eagle eye, so I joined another service and shipped out to the Pacific. I've often wondered if I would have had the right stuff to make it through such infernos as Saipan, Tarawa and Iwo Jima with a proud Marine regiment.

The Marines had invited me to Quantico to give a lecture on guerrilla warfare. I started with a critique of the U.S. Army's combat operations in Somalia, explaining how the Army brass had made the same mistakes as in Vietnam by employing firepower and technology against a fanatic guerrilla and failing to understand the enemy's culture, values, language, methods or motivation. Then I talked about the Vietnam War and gave examples of how successful Army units, such as the 9th Division's famous Hardcore Battalion, out-guerrilla'ed the Viet Cong, killing over 2,700 enemy soldiers at a cost of twenty-five American warriors.

Sadly, the Vietnam experience has been forgotten and the hard lessons paid for by the blood of 360,000 men have been lost because the U.S. armed forces never conducted a detached, in-depth study of that bad war. After Vietnam the Army went right back to refighting WWII, the kind of Desert Storm warfare the Military Industrial Complex so dearly loves, where firepower and the outpourings of our assembly lines, not skill, blow the enemy away.

Many lives will be saved because the Marine Corps is wisely sifting through the ashes of Vietnam, for as Somalia showed, that kind of low intensity combat - warrior versus warrior - will be the main event during the next several decades. Hopefully, the Corps' commitment to learning from the past will cause an ever rivalrous Army to follow in their wake.

But many of the young tigers I talked to are worried. Entrusted with the terrible responsibility of leading men in battle, they're rightfully concerned about having both the right stuff and enough training money to prepare their people for this ultimate and lethal Super Bowl. One captain said, "If we don't make our rifles, uniforms and boots big ticket enough, we'll end up fighting the next war with the same stuff our dads used in Vietnam."

They can't understand how the Pentagon and Congress can continue to buy high-tech rip-offs such as the two-billion-a-pop Stealth bombers, exotic and costly submarines and ships, and pour billions more into unneeded cold war hardware when their warriors will meet the enemy with worn-out, second-class gear.

These warriors are clued-in concerning pork, politics and history. They know their granddads fought at Guadalcanal with the recycled weapons their dads had used against the Kaiser. They can't blow the whistle, but you and I can demand that the brass and politicians get it right for once and stop sacrificing our best and brightest on their altar of greed and wrongheaded priorities.

**Used with permission of Colonel David H. Hackworth. This originally appeared August 9, 1994.**

# A MYTHICAL MARINE

Hung Hsiu-ch'uan had failed his examination for a post in the Chinese Imperial civil service. Today, in a Western nation, such a failure might easily be shrugged off, but for Hung it meant disaster. Unable to work for the Emperor, doomed to struggle through life as an impoverished Schoolmaster, he suffered a nervous breakdown. During this illness visions appeared to him. Interpreted in the light of some Christian tracts that he had been reading, these dreams convinced Hung that he was destined to end paganism in China. From his zealous preaching sprang the T'ai P'ing ("Heavenly Kingdom of Great Peace") rebellion, a bloody religious war which would claim millions of victims between 1848 and 1864.

As if rebellion were not enough, the Chinese Empire soon found itself at odds with France and Britain. Opium was the cause of the conflict, as the Chinese attempted to halt British traffic in the drug. From the head of the house of Manchu to the lowliest peasant, every Chinese scorned the Westerners and hated their "inferior" customs - and naturally there were numerous clashes between Chinese and foreigners. Early in February 1856 a French missionary was condemned to death by a Chinese court, clearly a case of legalized murder.

In October of the same year, the Chinese crew of a small British vessel was arbitrarily arrested and jailed in defiance of the British flag. Both European nations now were determined to punish China as soon as they could muster enough troops. In the meantime their naval vessels began sporadic combat operations along the China coast, operations which later became known as the Second Opium War.

In Canton, one of the five ports in which Westerners were allowed to trade, anti-foreign feeling was running high. Because of the perverted Christianity of Hung's militant disciples, missionaries were looked upon as spies. Traders also were despised for the merchant, even if he did not stoop to traffic in opium, was engaged in what the Chinese ruling classes considered to be among the basest of human activities. From this seething caldron of hatred, the American Consul at Canton called out for protection to Commander Andrew H. Foote of the twenty-two-gun sloop *Portsmouth* then lying eight miles downriver at Whampoa.

Early on the morning of 23 October 1856 five officers and seventy-eight men, among them Second Lieutenant William W. Kirkland and his eighteen Marines, rowed briskly ashore. This little force was organized into companies and posted on the housetops and in some newly constructed fortifications around the American compound in the city - but they seemed too few for the job at hand. The twenty-gun *Levant*, another sloop, dropped anchor at Whampoa on 27 October and added a score of Marines under Second Lieutenant Henry B. Tyler, plus a detachment of sailors, to the force already ashore. These sentinels exchanged shots with Chinese soldiers on 3 November, but no one was hurt. Captain James Armstrong, flying the flag of Commodore, East India Squadron aboard the thirteen-gun steam warship *San Jacinto,* arrived from

Shanghai on 12 November to assume responsibility for the protection of American nationals at Canton. Two days later he dispatched Brevet Captain John D. Simms and twenty-eight Marines to the turbulent city. Simms was placed in command of the entire force, including bluejackets.

Sustaining a garrison, even a small one, at Canton was a difficult job. From the diplomatic point of view, the presence of an American force in the midst of a fast-developing war could be taken as an insult by the sensitive Chinese. From a military standpoint, things were no better. Canton was located at the apex of a sprawling delta. Guarding the tortuous ship channel up the Pearl River were four forts located midway between the squadron's anchorage at Whampoa and the city of Canton, each of them incorporating the latest recommendations of European military engineers. Both Foote and Armstrong were keenly aware of the problem posed by the forts. To supply a garrison in the face of Chinese opposition would entail either running the forts or trying to slip past them in small boats at night. Either choice might involve the Americans in what was in reality an Anglo-French quarrel with the Chinese.

A decision on the part of Chinese officials to guarantee the safety of American interests at Canton brought a temporary respite for Captain Armstrong. Gladly he withdrew the bulk of the landing force, leaving only a handful of Marines at the American compound. In place of direct action, Armstrong devised an interim plan whereby *San Jacinto* and *Portsmouth* would wait downstream while *Levant* would hover off Canton in case the lives of the

Americans in the city should be threatened - but events were to intervene before this plan could be put into effect.

Whatever the intention of the Chinese in charge, they could not stem the rising tide of hatred. On 15 November, the very day the assurance of protection had been made, and while Foote was in the process of bringing the landing force back to Whampoa, the largest of the Chinese forts fired on the American boats. The next morning an unarmed boat from *San Jacinto* ventured to within half a mile of the fort farthest downstream. Captain Armstrong had dispatched the fragile craft to sound out a channel in case it became necessary to dash upstream. Without warning, one of the forts opened fire with both round shot and grape. The first volley screamed over the men crouched in the boat. Again the Chinese cannon roared in hate. Grape harmlessly churned the muddy water astern, but a shot crashed into the boat, killing the coxswain. A third salvo fell short.

Outraged at what seemed to be a deliberate breach of faith, both Foote and Armstrong decided to avenge this insult to the American flag. The more cautious of the pair was the squadron commander, Captain Armstrong. He hoped to cow the Chinese by engaging these so-called "Barrier Forts" with the guns of his ships. Since *San Jacinto* drew too much water to steam further upstream Armstrong transferred his flag to *Portsmouth*, and at 1500 on the afternoon of 16 November he ordered the expedition to get underway. A pair of small American merchant steamships, *Kumfa* and *Willamette*, battled the swift current to tow the sloops within range, however *Levant* ran aground before her guns could be brought into play so *Portsmouth* continued alone. At 1530 the Chinese unleashed their first salvo, and the Americans replied. As long as there was enough daylight to aim, cannoneers blazed away. Although several shots pierced *Portsmouth's* hull, while grape played havoc with her rigging, her only casualty was one Marine seriously wounded. In all, the vessel had fired 230 shells plus grape shot during the engagement.

A three-day lull followed as the Americans refloated *Levant* and repaired minor damage to *Portsmouth*. Armstrong began negotiations with the Chinese, but before he accomplished anything his health broke down and he turned command of the expedition over to the daring Foote - but before returning to *San Jacinto* the Captain advised Foote to withhold his fire unless the Chinese should attack.

Once Armstrong had left, the junior officer took stock of the situation. Facing him were four massive granite fortifications with walls seven feet thick and a total of 176 guns, some of them of ten-inch caliber, which could be brought to bear against an attacking fleet. In addition, there were rumored to be between five thousand and fifteen thousand Chinese troops in the Canton area. Although the forts were powerful, the strongest in the Empire, Foote need not fear the army, a poorly equipped, half-trained rabble. When Armstrong on 19 November ordered Foote to take any action necessary to forestall a Chinese attack, the Commander decided to seize and level their works.

Abandoning the idea of passive defense, the Americans now planned to head off a major battle by striking first - and on the morning of 20 November *Portsmouth Levant* went into action against two of the forts. Under cover of the ships' guns, a storming party of 287 officers and men, led by Foote himself, landed unopposed. Spearheading this force were the squadron's Marines, approximately fifty in number, under Captain Simms and a small detachment of sailors. Because of the terrain and the sheer walls of the first fort, the Americans had to assault from the rear. A village in which a handful of Chinese snipers had been posted loomed in their path, but the Marines quickly cleared the place and began the final sprint toward the redoubt. The defenders bolted, and some of them even tried to swim the river. From the captured parapet, a hail of American bullets cut into the fleeing horde. Some forty to fifty Chinese were killed.

At Canton, four miles distant, lay the main body of the Chinese force. No sooner had the stampeded garrison reached the city than an expedition got underway to recapture the first fort. While the fresh Chinese troops were approaching, Simms and his Marines had returned to the village just outside the walls to scatter a band of die-hards who had rallied there. A brisk volley and a fierce charge sent the enemy wallowing toward safety in the rice paddies. The Marines followed until the going got too difficult, paused to regroup, and began falling back. Suddenly the battered Chinese, their spirits revived by the coming of reinforcements, turned tiger and launched a counterattack. Well over a thousand men swarmed through the ooze of the rice paddies to engulf the Leathernecks. Simms had his men hold their fire until the Chinese were within two hundred yards. Then volley after volley thudded into the enemy ranks. Gamely the Chinese stood their ground and returned the fire, but Marine marksmanship proved too accurate and the enemy ran. Two other counterattacks were attempted, but both were beaten back by Leatherneck muskets and boat howitzers.

Scheduled for assault the following day was the second of the Barrier Forts. Early that morning, the Marines and sailors of the landing force piled into boats and, towed by the steam tug *Kumfa*, began moving upstream toward the objective. American guns lashed out above them in support of the landing, and the three works still in Chinese hands divided their fire between the pair of sloops and the line of boats. A sixty-eight pound shot knifed through one of the American boats, killing three and wounding five - yet the enemy's fire, though frightening in volume, was for the most part inaccurate. Once ashore, Simms led his men across a creek waist-deep with murky water and over the granite walls. While a force of a thousand Chinese hovered just out of range of the tiny American howitzers, Corporal William McDougal of *Levant* planted the Stars and Stripes on the parapet.

Once the fort had fallen, Foote ordered Sims to clear the Chinese from the riverbank so that his boats would not be caught in a crossfire during the next phase of the operation - an attack upon an island bastion in the Pearl River. Hugging an embankment, the Leathernecks were moving cautiously forward when they collided with a Chinese battery of seven guns. Caught completely by surprise, the enemy fled amid a fusillade of musket fire. Leaving a handful of men to destroy the guns and protect his rear, Simms moved his force to the top of the embankment and opened fire across the water to silence the third of the Chinese works. Once the guns of the island fortress had been stilled, Simms and his Marines withdrew along the embankment to join in Foote's next assault.

This third fort fell quickly to the American assault force. Fire from the two captured citadels and from the opposite shoreline blanketed the works in a shroud of dust and smoke. Once again, Corporal McDougal broke out the American flag as the assault wave surged over the walls. On the second day of the operation, 21 November, two forts and a Chinese battery had been taken. All that remained was to capture and destroy the last of the works, Center Fort, on the Canton side of the river.

Preparations for this final phase began in the darkness of the following morning. All captured artillery pieces which could not be used to support the attack had been torn from their mounts and spiked, but the best of the Chinese weapons were aimed at the squat heap of granite that was Center Fort. The sky was barely light when an American howitzer snarled across the water. The enemy did not reply. Again the cannoneers tried to draw Chinese fire, but there was no answer. Then three waves of boats crawled out from the island toward the final objective. The howitzers and captured cannon roared in support of the assault waves, but Center Fort remained quiet. All three lines of bobbing boats were well within range when the Chinese at last cut loose. Clouds of grape shot whined across the river as the men of the assault force leaped into waist deep water and began wading toward the base of the walls. Once they had clambered to the parapet, they found that the enemy had fled. A crude sort of booby trap, a cannon loaded and aimed at the boats, had been left behind by the defenders - but alert Marines quickly snuffed out the smoldering powder train.

Since Foote's squadron now was in complete control of the barrier fortifications, the work of destruction could begin in earnest. Those guns which had been spared to assist the final assault were uprooted and spiked. The ruined pieces then were rolled into the water. Demolition parties moved from fort to fort planting charges of gunpowder beneath the mighty walls. On 5 December a spark believed caused when someone's crowbar glanced off the granite touched off the powder being placed beneath the walls of Center Fort. The blast killed three men outright, and wounded nine others. On the following day the two sloops moved downstream to their normal anchorage at Whampoa, and behind them the most formidable works in the Chinese Empire lay in ruins.

In one brief but furious campaign, Commander Foote's command had captured four powerful redoubts, killed an estimated five hundred Chinese, and routed an army of thousands - all at the cost of seven killed in action, three killed during the demolition of Center Fort, and a total of thirty-two wounded or injured. None of the Marines were killed in the fighting but one, Private William Cuddy, took sick and died, while six others were wounded. Three days of the fiercest action proved that ships, when teamed with a strong landing force, could indeed fight forts.

Besides being a truly remarkable feat of arms, the destruction of the Barrier Forts appeared to be a diplomatic success when an apology for the unprovoked attack of 16 November on the sounding boat was quick in coming. Foote had avenged an insult to the American flag and made certain that the Chinese at Canton would behave in the future.

The men of the East India Squadron were justly proud of their achievement. To commemorate their comrades killed at Canton, they raised a thousand dollars to erect some sort of monument. Nothing, however, was done until Foote detached from *Portsmouth* and arrived at the Brooklyn Navy Yard in October of 1858. As Executive Officer of the Yard, he was able to begin work on the monument. The site selected was just inside the Sand Street Gate, and under the hand of a local sculptor a marble shaft surmounted by an eagle gradually took shape. At its base was a tablet listing the names of those who fell in the attack.

When it was dedicated late in 1858, the marker listed twelve names: E. C. Mullen, Louis Hetzell, Thomas Crouse, James Hoagland, William Mackin, Alfred Turner, Edward Riley, Joseph Gibbings, Edward Hughes, Charles Beam, Thomas McCann - all sailors - and one "John McBride - Marine." Unfortunately, there were errors on this roll of honor. The names of Lewis Hetzel and Thomas Krouse had suffered at the hands of the stonecutter. Worse yet, there was no sailor named Thomas McCann killed at Canton, nor was any Marine killed during the battle. Who, then, was John McBride? None of the Marine detachments involved in the action carried anyone by that name on their muster rolls. Eager to finish the task, the impetuous Foote apparently had not taken time to check official records. Like Thomas McCann, McBride was the result of a lapse of memory.

Captain Armstrong, Commander Foote, and Brigadier General Commandant Archibald Henderson had all hailed the exploits of the sailors and Marines of the East India Squadron. The Secretary of the Navy, in his Annual Report for 1857, had devoted an entire paragraph to the Battle of the Barrier Forts. It is ironic indeed that the memorial to Foote's gallant dead, a work which he himself began, should contain not only misspellings - but the names of a phantom sailor, and a mythical Marine.

# PROUD SON OF A MARINE

John S. Wallschlaeger

My father was a United States Marine. His name was Edward Wallschlaeger, a Parris Island graduate at the age of seventeen, in the Class of August, 1946.

He had many stories about his time in the Marines. Some were funny, some were sad, but all were worth listening to. Being a Marine sometimes got him into trouble, and other times it kept him out of it. One story he told was about an incident at a tavern in Wisconsin. After grabbing an armful of shot glasses to use as projectiles to get out of the bar, the group he was with returned to the cabin that they had rented for the weekend. As my dad told it, the other half of the disagreement was not far behind them, and the biggest one in the bunch pounded on the door of the cabin. My dad opened the door. The guys outside were yelling for them to come out. The big guy stepped forward to grab the door and noticed my dad was wearing his Marine Corps issued T-shirt. He asked if he was a Marine, at which he replied, 'Yeah, you want to make something of it?" Seems the guy was Army, and had been in Korea. He said, "You Marines saved my unit's ass over there." He turned to his buddies and said, "Come on guys, we got no axe to grind here," and they left. Clearly being a Marine has all kinds of advantages.

My father died in July, 1997. In as much as I recall him speaking very fondly of his time in the Corps, I never really appreciated what it all meant until the Marines helped our family bury him. Their presence and assistance exemplified the meaning of "Semper Fi." I am, and will always be, forever indebted to the Corps for making my father a Marine and for helping us bury him, and I see now how his being a Marine has affected me. A few years back the Corps' theme was "The Change is Forever." Without a doubt it was true for my father, and it has also been true for me. While I did not serve in the military, I did serve many years under my father's orders and experienced his esprit de corp. We didn't have any yellow footprints at our house, but we most certainly had to toe a line or two in our days. I learned expressions like "squared away," and "burning daylight." My father was forever utilizing what he learned in the Corps. He taught us boys firearms safety and the art of improvising, adapting, and overcoming. He could be seen field-stripping his Phillip Morris whenever needed, and on occasion was seen putting his cigarette ash in the cuff of his pants when no other option was available. When doing chores outside or hiking or hunting, his K-Bar was always nearby. He would always look at it with a sentimental eye as he would put it back in its sheath. I now have and cherish that K-Bar. I could go on and you would read what I imagine is typical of most Marines - love and respect of God, Country and Corps. He instilled those values, along with Honor, Courage and Commitment, in my siblings... but especially in me. His tutelage helped make the change in me last forever.

Yes, the Marines helped us bury our father. I am grateful to the Marines from Marine Wing Support Squadron 471,

Detachment Bravo, of Green Bay Wisconsin who served as the Honor Guard for his visitation and burial. I never mentioned anything to my father before he died about the Honor Guard idea. I knew he would humbly decline. Yet my mom, a loyal Marine wife of over forty-six years, and I knew better. In fact, my mom loved the idea. She was also grateful for the Marines who tended to his funeral, and recalls fondly the days when my dad was in the Corps. She specifically cherishes her trip to Washington D.C. in the spring of 1951 to see her Marine on leave from Camp Lejeune. They knew he could be sent to Korea, but those orders never came and they got married in November of 1951. As my mom looked at the Marines in July 1997, I know she found comfort in their presence. I did too. I also found it ironic. One of the things my dad told me about of his time in the Corps was that he had served on several Honor Guard details. He helped lay to rest Marines who gave their all in Korea. He described it as very honorable duty, but difficult duty as well. Sergeant Barber, who led the detail, told me that for Marines the idea of being there is simple. During his time my father was there for the Marines, and now it was the Corps' time to be there for him. Gosh were they ever. When the Marines shook my hand and told me how sorry they were that my father had passed on, it was as if they had served side by side with him at Lejeune or Cherry Point. The image still blurs my vision and moistens my cheeks.

In between the time I arranged for the Honor Guard with Sergeant Barber and the burial, I did some volunteer time at the local County Fair in a Crime Stoppers tip line tent. As fate would have it, next door to our tent was a Marine Corps Recruiting tent, pull-up bar and all. Already grateful for the Corps' assistance with the upcoming funeral, I went over to the Marine working the tent to thank him. Staff Sergeant John Jamison of Recruiting Station Milwaukee was on deck. He gave me a very warm reception. He even challenged me to do twenty pull-ups for a T-shirt. With my dad in mind, and pulling a few muscles I didn't even know I had, I got that twenty - and the shirt!

The Marines who attended my father's funeral were very squared away, with fresh high and tights and sharply pressed uniforms across the board. No detail was overlooked. After the detail folded the flag at the cemetery the Gunny presented it to my mom. Contained inside of it were some spent rifle casing from rounds they had fired in his honor (the funeral home wasn't too keen on cutting paperwork to allow for it at the cemetery). Gunny, as he presented the flag, said some kind words about the era of Marine my father served with. My mom genuinely appreciated those words. One of the Marines blew taps on a bugle, and we said our final goodbye. We later had a luncheon which the Marines attended. The brotherhood was instantaneous. Many of the other people in attendance, friends of my parents, had served in other branches of service during WWII and Korea. The Marines could hardly get a sip of their refreshment without someone coming over to them to shake their hands and thank them for what they did. It was really something to admire from a distance. I recall one of my father's friends saying "Leave it to the Marines to do it right."

A few months after the funeral I heard from Staff Sergeant Jamison, who asked if I would like an opportunity to see how a Marine is made. He said he had a slot open in an upcoming Educators trip to San Diego. Of course I couldn't answer 'yes' quick enough! I made the arrangements at work (and my boss asked me not to enlist while I was there) and in February of 1998 I was off on a Navy DC9 to MCRD. My shuttle ride to and from Milwaukee Mitchell Field was provided by a recruiter in the area, and he was a darn squared away Marine with lots of ribbons and a barrel chest to pin them on. I found myself pinching myself. Could it really be that I was off to MCRD?

On the way to San Diego I got to know a few people on the plane. One was a Fire Chief in the Milwaukee area and a RVN Marine, with two tours. He spoke highly of the M60 he carried there, had USMC related items in his office, and told me he never had a problem with any Marine he had hired as a firefighter for his department.

The day after we arrived we rode by bus to the Recruit Depot. When we got there a DI boarded and asked all Marines to get off, and once they were off he stepped back on. Then, I am sure in much the same tone he gives to the recruits who arrive there early in the morning hours, he began his welcome speech - or should I say *shout*. I was videoing it, but his demeanor was so intense I began to wonder if he wasn't going to ask me what the hell I was smiling about and yank it out of my hands. He made it clear it was time for us to get off "his bus" and get out on those yellow footprints. At this point I think it would have taken wild horses to get the grin off my face as I was having the time of my life. Mind you I wasn't seventeen or eighteen years old, and knew this was a drill. When we got on the footprints a couple of other DIs joined in on the fun and proceeded to get in our faces. Well, I darn near peed my pants! When it was done we received a card signifying we were honorary Marines for having stood on the hallowed "yellow footprints" for two minutes. Two minutes! Is that all it was?

Our workshop included a complete tour of MCRD, and we saw recruits in nearly all phases of the indoctrination. We saw them marching, we saw them running, we saw them swimming, we saw them getting, well - indoctrinated. We went out to Camp Pendleton and toured the "Crucible" stations. We ate some Marine Corps chow, and visited with some recruits. To top it off we were privileged to see a graduation. There wasn't a dry eye in the reviewing stand. To say it was enlightening is an understatement. I now understand what it means to be a Marine, and why it is such a brotherhood (with no offense intended to any woman Marines).

Our flight back was pretty normal, but there was no getting my father and the Marine Corps out of my mind. Staff Sergeant George, the recruiter who had taken me to the airport, was there waiting when our plane got in despite the fact it was delayed and very late. A new friendship was forming, and I stayed in regular contact with him and had opportunities to help him recruit a few Marines. Scott also had me give a speech to a group of delayed entry recruits and their families about what being a part of the Marine Corps family meant to me. It was an honor to be a part of the process.

Scott and his wife Michelle invited my wife and me to a Marine Corps Ball in November, 1998. How honored I was to be there. Prior to the Ball dinner Scott re-upped for another tour, and again it was clear to me that I was a very lucky individual to be a part of all this. Over the years we have stayed in touch with the Georges as the Marine Corps life moved him around the country and the world. He was in towns which were not on any maps I had, and they kept us informed as to his deployments.

I had a standing invitation from Scott, now a Gunnery Sergeant, to come and visit him at MCAS Cherry Point in North Carolina. In February 2003, my six-year-old son Oliver and I took Scott up on that offer and flew down to see him and his family. Scott was now a short-timer and would be retiring from the Corps later on in 2003. We had a great time on our trip. We got to see what was left of the Marines and equipment stationed at Cherry Point, and we also went to Camp Lejeune and saw what we were allowed to see there too. The ramp-up for the war in Iraq had nearly emptied both bases. It occurred to me that I was seeing something in real terms which most others would not get to see, the effect of preparing for war. My son asked me when he was going to see all the Marines, and I told him the few we saw around the base were it and that the rest had been shipped off to the war. At age six that didn't make the same impression on him as it did on me. It made me pause to imagine what it means to the children of those Marines who had shipped out.

We took photos of buildings and locations which were nearly identical to pictures my dad had taken back in 1950. That was eerie, being that it was more than fifty years later! We posed for Scott on top of tanks on static display, and in front of the new MTRV seven-ton truck which is made in Oshkosh, not far from where we live. The Gunny took us to the PX and made sure Oliver had his cammies for the trip back home, in addition to a set for his younger brother. Of course, Oliver saw to it he had the rank of Gunny on his collar, and Scott even decorated him with a few ribbons. By the end of our visit the smile on my son's face was almost non-stop. Once again it occurred to me that the change was continuing to be forever.

A day does not go by that I don't think of my father and the Corps. The Eagle Globe and Anchor can be seen flying from our flagpole, stickered on our vehicles and woven into the fabric of our clothes. The Marines' Hymn is even the exit sound when I shut down my laptop. The only thing more prominent around our home is the American Flag. I recognize and appreciate the term 'honor' more every day.

So as I close let me refer back to the expression, "The change is forever." I think I have detailed that for my father it certainly was, and I believe the change the Marine Corps has instilled in me will also be everlasting. Now let me add that forever is a long time, so I am doing my part to keep the spirit alive. My two sons love wearing their cammies, saluting the flag (covered of course) and can often be heard humming the Marines' Hymn. They know being a Marine is something special. Many of these things have been rooted in me, courtesy of my dad - Marine Corps class of 1946. So will the change continue on in my sons too? Time will tell, but I sure hope so. What I do know is we are off to a very good start.

# REPORTERS FIRST
## *Americans Second*

*All I can say is, with people like Katie Couric and Chris Mathews in the news business, not much has changed:*

In a future war involving U.S. troops, what would a television reporter do if he learned the enemy troops with which he was traveling were about to launch a surprise attack on an American unit? That's just the question Harvard University professor Charles Ogletree Jr., as moderator of PBS' *Ethics in America* series, posed to ABC anchor Peter Jennings and 60 Minutes correspondent Mike Wallace. Both agreed getting ambush footage for the evening news would come before warning the U.S. troops.

For the installment on battlefield ethics Ogletree set up a theoretical war between the fictional "North Kosanese" and the U.S.-supported "South Kosanese." At first Jennings responded, "If I was with a North Kosanese unit that came upon Americans, I think I personally would do what I could to warn the Americans."

Wallace countered that other reporters, including himself, "would regard it simply as another story that they are there to cover." Jennings' position bewildered Wallace. "I'm a little bit of a loss to understand why, just because you are an American, you would not have covered that story."

"Don't you have a higher duty as an American citizen to

do all you can to save the lives of soldiers rather than this journalistic ethic of reporting fact?" Ogletree asked. Without hesitating Wallace responded, "No, you don't have a higher duty... you're a reporter." This convinced Jennings, who conceded, "I think he's right too, I chickened out."

Ogletree turned to Brent Scrowcroft, who later became the National Security Adviser, who argued, "You're Americans first, and you're journalists second." Wallace was mystified by the concept, wondering "What in the world is wrong with photographing this attack by North Kosanese on American soldiers?" Retired General William Westmoreland then pointed out that "it would be repugnant to the American listening public to see on film an ambush of an American platoon by our national enemy."

A few minutes later Ogletree noted the "venomous reaction" from George Connell, a Marine Corps Colonel. "I feel utter contempt. Imagine two days later they're both walking off my hilltop, they're two hundred yards away and they get ambushed. And they're lying there wounded. And they're going to expect I'm going to send Marines up there to get them. They're just journalists, they're not Americans."

Wallace and Jennings agreed, "It's a fair reaction." The discussion concludes as Connell says, "But I'll do it. And that's what makes me so contemptuous of them. And Marines will die, while going to get a couple of journalists."

**From the April 1989 *MediaWatch*, a monthly newsletter then published by the MRC.**

# RETURN TO IWO JIMA

Steven McCloud

*I couldn't help but think of my friend John Butler when I first read this. John was himself a Captain of Marines, while his Father - Lieutenant Colonel John Butler - was KIA on Iwo while serving as CO of 1/27. A few years ago John and his brother visited the very spot where their Dad fell, and I can't imagine what that must have been like for them.*

It was 0300 and the entire group of about a hundred and thirty of us had once again assembled in the lobby of our hotel, the Outrigger, on Guam. My own eyes were dry and burning because I'd stayed up too late with the guys as we discussed our days on Saipan and Tinian.

All around me were the now familiar faces of Major General Fred Haynes and Lieutenant General Larry Snowden, Colonel Tom Fields, Medal of Honor recipient Jack Lucas, and my own friend Danny "Doc" Thomas who was a Corpsman with the Fourth Marine Division in 1945.

Standing a few feet away was Historian Stephen Ambrose, making his first venture into the Pacific. There was James Bradley, whose recent book had done so much to return the focus to the Pacific Theater. Sitting nearby was Pamela Marvin, wife of the late Lee Marvin, the actor and Marine who was wounded on Saipan. All around I saw the faces of the unforgettable personalities who had dreamed of making this journey into the Pacific, and today we would stand on Iwo Jima.

At three o'clock in the morning there's only so much that can go through one's mind. I vividly recall on that morning, my own seemed able to generate only one recurring thought - *Iwo Jima? Iwo Jima. Holy cow.* I would shake my head and snicker silently with disbelief. It was the last little part that I recall so clearly. It was a realization that this was serious. Everyone in that outstanding group held the highest reverence for the battle sites we visited and for the men who fought and suffered there. It was that reverence which gave me such admiration for this group of citizens who wanted to go to Iwo Jima. Indeed, there was a bond among us all. I've never felt so comfortable - but only on this morning did the sobering realization find me. This was not simply a visit to another battlefield. This was, for all of us, the pinnacle.

As we flew above the clouds that morning I sat and chatted quietly with Mrs. Lee Marvin. I was inspired by her soaring compliments of her husband and the Marines with whom he served. Even so there was a somber mood on the plane during that morning flight.

When we boarded our two 737s there was soft, soothing music playing and images of island waterfalls on the small screens above the seats. Continental Airlines had placed a small bag on the seat for each person. In it was a ball cap with stitching in the back which read, "Iwo Jima Reunion, 2001." There was also a small glass cigar tube printed with, "Sands of Iwo Jima, Courtesy of Continental Airlines." This was not just another charter for Continental.

As we neared the island, only the whine of the engines could be heard. That whine lowered to a soft hum as we began to descend nearer and nearer to the white layer beneath us. The Captain quietly informed us that we were about "seven and a half miles out from Iwo Jima." No one spoke. I'm sure the veterans were lost in their memories of the island as they last saw it, and of the buddies they last saw there. Many of them had left their youthfulness on that island and come away with horrific nightmares. Some of them had lived with shame and guilt that they had survived while their buddies had not. Some of them were afraid of what the sight of that black island might call forth from the darkest chambers of their memories. That reality was only moments away.

I watched the flat cottony layer of clouds come nearer and nearer as I imagined the Japanese fighters which had flown these skies decades earlier, when they were still in existence. Mostly I pictured the island of Iwo Jima, and all that I knew it to be. It was waiting for us, just beneath those clouds. Then I watched as we settled slowly onto that cottony layer and as the wing disappeared.

The whine of the 737's engines fell suddenly quiet as we descended into the clouds. I sat alone, peering out my window, watching the plane's silver wing disappear until I could see only white. In those few moments I thought about how strange this all was. It was as though these two sparkling airliners were taking us on a journey back in time to a lost world, a place of legend. The very name of Iwo Jima had taken on a life and meaning all its own, one with no synonym.

I counted the seconds while watching for a glimpse of land beneath us, lest we emerge from the clouds directly over the island. At last, after perhaps ten eternal seconds, I saw darkness through the clouds and quickly recognized the dark Pacific. I stooped my head to more quickly see underneath the cloud cover, and, looking ahead to the horizon I saw it just off to starboard. My disbelief emerged in a whisper. "Iwo Jima."

We were approaching from due south of Mt. Suribachi. It was a dark, overcast morning and all around the southern approaches to the landing beaches the black ocean was pierced by small, distinct, almost blinding patches of silver sunlight. It was startling, and to my eyes it may as well have been the fleet lying off Iwo Jima preparing for the landing.

Slowly we eased around the southern tip of Suribachi. The island was no longer a mere photograph. Our pilot hung the plane on its engines, flying more slowly than I believed a 737 could and still remain in the air. For half an hour we circled the island in the air as he dipped the wing so we could all see this place. All the while I kept wondering what else in this world would possibly elicit such an emotional response from ordinary people. What was the lure? What was it about this place? I can still visualize our pass just above the crater of Suribachi. I could easily imagine the thousands of Marines on that island. These American Marines had come here some twenty years before I was born. Indeed, my father was merely five years old when this had all taken place - and yet it was real to me and to all who were on this plane. It was real to the Continental crew. The pilot circled repeatedly so that people on both sides of the plane could see. He knew the significance of this place. A friend who was already on the ground in the lead plane later told me, "I thought you guys were never going to land."

The island is now largely covered with green growth. No longer does it show the scarred and charred black face that it did in 1945 after seventy-two days of bombardment, but one can still see the incredibly broken terrain against which our troops threw themselves. The plateau running through the center of the island bows much more dramatically than I had imagined. What's left of the "Meatgrinder" is densely green in patches, and I could not help but wonder what lay beneath that scrub brush.

Then, as we continued to circle, I came close to all which had happened on this island, the unthinkable sacrifice of youth and innocence, the unimaginable selflessness and heroism, the tragic terror cast upon all those young men. It can still be seen from the air. The tears came, and would not leave. The ground came nearer. I came closer to touching down on Iwo Jima, and the tears would not stop. Then I felt the wheels touch lightly onto the runway and all at once I was back here with all of them. I knew the names of so many, and yet so few. Jerry Gass, "John Q.," Ed Rickets, Bahnken, Willis, Thomas, Haddad. So many of my friends had been here when this was Hell. So many others had ceased to exist after setting foot on this island. Now I was on Iwo, and I was afraid I might not be able to get off the plane. I kept whispering to myself, "you blubbering fool."

I managed to compose myself long enough to gather my camera bag and move toward the center aisle, where Pamela Marvin was waiting for me. She smiled and said, "Congratulations." I couldn't utter a sound. I bit my lip, nodded, and turned away. I had not cried like that in years.

The veterans exited the plane first, then family of the veterans, followed by us tag-alongs. The outstanding Marines of the 31st MEU greeted us all with two welcome lines. The pride on the smiling faces of these young Americans was unmistakable. They were proud to be on Iwo Jima with the men who etched the name into history - and so was I.

After a quick briefing near the hanger, we all mounted the Marine vehicles to begin our day on Iwo Jima. I was already in one of the five-tons watching the Marines help Cyril O'Brien, the tough ole' feisty, five-foot six-inch veteran Marine Combat Correspondent, up the ladder and into the truck. After a few grunts and groans he found his seat and proclaimed as only a Marine can, "Getting' old's a real pain in the ass!"

I ended up in a Marine Humvee with the Spielberg film crew. They were filming every step of my friend Danny Thomas' return to the island. Danny was a young Corpsman with the 2nd Battalion, 23rd Marines, 4th Marine Division at the time of Iwo Jima. He became a Navy Corpsman to stay out of field combat, visualizing instead a job in some clean, sterile Naval Hospital, but soon found himself assigned to the "Fleet Marine Force," whatever that was. He had no idea. He'd lived with horrific nightmares for fifty-six years, too terrible to describe here. Somehow he had mustered the courage to return to the island. They filmed every minute of it.

As we left the airfield Pat Mooney, Executive Vice President of Military Historical Tours, pointed behind me. "There's the Amphitheater, Steve." I wanted badly to go there later, but we were off to the Ceremony Area. They had staggered our planes by thirty minutes to aid in logistics on the island, so while we in the second plane were still in the air the passengers from the first plane were transported up Mt. Suribachi for their early morning view. When our plane arrived the mission was repeated, so that everyone could visit that hallowed place before the Joint Commemorative Ceremony.

Both American and Japanese veterans of the battle were present and participated in the solemn ceremony. The remarks were brief and poignant. The ceremony was serene. Then came Taps. Those first three beautiful notes coming from my left sent chills through my body... but then the same notes were echoed by a bugler from somewhere off to the right. It was too much. As tears streamed down my cheeks I could only close my eyes and inhale their cry for the fallen. "I will never forget this," I thought. Another friend on the tour was across from me, some fifty feet away. He later told me, "When he started playing Taps I was okay, but when the other Marine started echoing him, I lost it."

After a packed lunch handed out by the Marines, it was every man for himself. The 31st Marine Expeditionary Unit, based on Okinawa, had landed in advance, bringing all necessary facilities for the occasion including numerous heavy trucks and Humvees. They maintained a constant circuit around the island and we could catch a ride to any point we desired, get off where we wanted, and hitch another ride later. As for me, I wanted to get to the beach.

I struggled through the coarse black sand from Blue Two to Yellow Two, where my friends in Fox Company, 23d Marines had landed. I followed their advance toward Airfield Number One and headed for the "turntable" where they spent their first night. I was on the beach for two hours, and saw not a soul in either direction. It was a strange encounter with Iwo Jima. I stood on the beaches once cluttered by the debris of war and sixty thousand Marines, always with Surbachi looming in the background. Now here I stood, alone with Suribachi. It was as though I was alone on Iwo Jima, one of the most infamous places on Earth. I had it all to myself.

Negotiating one's way through the sands of Iwo Jima still requires a great deal of effort. Each step crunches loudly, and leaves a large depression. The ash is actually more like very small black gravel than sand. Any of that in your boondockers could cause a serious problem in short order, but the difficulty today is nothing close to that experienced by the Marines who struggled through it in 1945 carrying sixty to seventy pounds or more of gear, weapons and ammunition.

Today the beach is packed tightly, whereas in 1945 it had all been loosened by continuous bombardment. Naval gunfire, enemy mortars and artillery, and American bombs had all turned the beach into a quagmire which served to "fix" the American forces in place for Japanese fire. As I slogged through this crunching sand I imagined how difficult it must have been to move inland, over the terraces and up the steep rise toward the airfield, all the while taking unrelenting fire from everywhere.

Meanwhile, near the base of Suribachi, a surprising discussion had developed. Phil Mongillo had been a Corpsman with the 1st Battalion, 28th Marines on Iwo Jima. Now in his mid to late seventies, Phil is a marathon runner. I was told that a few years ago he had surgery for cancer, and two weeks later ran the New York Marathon. That's the word. Just as a truck was about to carry a load of visitors to the top of Suribachi, Phil stopped the Colonel. "Colonel, I have my own way in which I would like to honor the Marines who died here. If you don't see a problem with it, I would like to run up Mt. Suribachi instead of riding." The Colonel replied that he didn't have a problem with it, and immediately a young Marine stepped forward. "Sir, I would like to run up Mt. Suribachi with you." Another Marine stepped forward, "Sir, I would like to run with you as well." Another followed.

Phil was stunned, but certainly pleased. It was fine with him. He suddenly had a group of several Marines, and moments later the entire group of Marines and one old veteran in his seventies were jogging up the steep switchback road to the top of Suribachi - he in his running shoes and shorts, they in their boots and BDU's. Once they reached the summit they had a look around, made photographs, looked at each other and agreed, "Well, let's run back down." This time the group was even larger. Semper Fidelis.

Bill Hudson was a BAR man in the 4th Marine Division in March of 1945. On 14 March 2001 he met a Japanese veteran who wore a red cap which read in Kanji, "Iwo Jima." The Japanese gentleman had manned a machinegun in a pillbox on Yellow Beach and was captured after being wounded. Bill met and spoke with him through an interpreter for a few moments. They made photographs together, shook hands, and exchanged hats, Bill's 4th Marine Division hat for the Japanese Veteran's Iwo Jima cap.

After our full day on the island the time came for us to leave. My friend Danny Thomas felt as though an enormous burden had been lifted from him. "I feel so much lighter," he grinned with a bound. That night back on Guam, he slept soundly without the usual nightmares. There have been no more of those horrible dreams since that night.

As we boarded the plane and taxied to the end of the runway, dusk had given way to darkness. Again, there was complete silence aboard the plane. I did not want to leave. As we roared along on our take-off run I gazed out at the blue lights, waiting for the moment when I would leave Iwo Jima. We were off. It occurred to me that they had literally "opened" the island for us to come on that one day. As our wheels left the ground, the island was once again closed.

*Return to Iwo Jima* **appeared in the July and August 2001 issues of the** *Always Faithful* **newsletter.**

# THE WOUND

**Dick Moskun**

**"It was a shot in the arm, which is better than a shot in the head."** - Richard Armour

*I sometimes get criticized for not including enough material about Marine Air and must conclude that, as a recon guy, it is because I have lived through a lot more takeoffs than landings. Hopefully the inclusion of the following story, which is "part of the collective history of VMO-3," will get me back into the good graces of the airdales.*

VMO-3, Phu Bai, Fall of 1967. A correspondent from *Stars and Stripes* came looking to interview "Mouse." The story was he had captured a North Vietnamese soldier in the field - a rare occurrence for a Huey gunship crew chief. His version was never published in *Stars and Stripes*. Mouse didn't have the ability to relate the facts of the incident without a lot of embarrassment and awkward jokes, so he directed the correspondent to "Saint," who had also taken part in the capture. This is Mouse's version - thirty-three years later.

The purpose of our single gunship mission was to be an observation platform, a containment element, and if necessary provide cover fire for a squad of Marines who were sweeping the length of a small peninsula jetting into a bay of the South China Sea. They were checking the locals and looking for targets of opportunity, which was a routine assignment for them.

After arriving on station it didn't take long for us to spot a twenty-foot canoe leaving the shore opposite from the sweeping Marines. Major S (the pilot) maneuvered the Huey near the canoe. We waved to the elderly couple, motioning them back towards the shore. They smiled and waved back. Major S circled and flew lower. I pointed to them, and then pointed back to the shore. The Mamasan shook her head "no." I told Major S, "I don't think they want to do it." The Major admonished me for taking our assignment too lightly. We circled around one more time. The Major asked me to fire a couple of rounds in front of the canoe with the door gun. I was surprised, and amazed, by the waterspouts the bullets made. They must have been ten feet tall. I also remember the expression on Mamasan's face. It wasn't fear or terror that I saw. It was more like anger, or great disappointment. I realized that maybe for the first time in-country, I had just shot towards an innocent. The canoe began to turn around. The Major flew us back down the middle of the peninsula toward the Marines.

We were zigzagging just over the trees when we spotted a bad guy. I think the maneuvering spooked him into running. Major S pulled up hard, swung to the right, and slid the Huey back down towards him. He appeared to throw away a rifle in some bushes, then quickly doubled backed under us. We circled for another run, and by now he was about halfway across a rice paddy heading for a thicker tree line. I suppose it was during this time Major S thought we could capture the

133

guy. The Major came in close and fast trying to knock him down with the skids (Huey's don't have wheels). It was great fun, but it didn't work. Splashing through the mud, the bad guy was getting nearer the trees. After another attempt at knocking him down on his keister, Major S slid the Huey to the ground within arm's reach of the bad guy and said to us, "Go get 'em." The bad guy was next to Saint, (the door gunner). Saint leaped from the Huey toward him. He would have gotten him too - if it weren't for the six-foot safety belt still attached to his waist. Saint doubled up and fell into the rice patty. It was a dramatic effort, but not a pretty sight.

Seeing him on the ground reminded me to unbuckle *my* belt and take off my helmet. By the time I jumped out of the cabin the bad guy had made it through the trees. Saint was up and running after him - and it was about then I realized we didn't have weapons - so I reached back for my M14 and tried to pull it out from underneath the bench seat. Most crew chiefs kept a mess of jumbled-up gear under there, and I tugged a couple of times but the rifle wasn't coming out. I reached for the next thing I saw, the survival hatchet - an angry looking tool about eighteen inches long. I ran towards the place in the tree line where Saint and the bad guy had disappeared, wondering all the while "what am I going to do with the hatchet?" Having made to the other side I saw Saint and the bad guy running along a small dike separating two fields. Saint might have been fifteen yards behind the bad guy, and I might have been another thirty behind Saint. It turns out we were now making great silhouettes for the squad of Marines - remember them? They must have been a hundred yards to our right. Mud began splashing around me, and I felt something hit my arm. I thought it was a rock, so I spun around to see who had thrown it. Then I saw the squad. They were in line, standing and kneeling, shooting at us. Like the Mamasan earlier in the canoe, I got pissed. Shouted some obscenities, flipped them off - but kept running. I knew they were shooting, but it never occurred to me they were shooting at Saint and me. We were young and dumb.

We had probably run seventy-five to a hundred yards, and Saint and the bad guy were slowing down. The Marines had hit the bad guy in the right side of his rib cage, but the forty-pound standard issue aviation flack vest Saint was wearing began taking its toll and he was no longer closing the gap. I had made up most of distance by running hard and carrying less weight, since I only wore the front plate of the flack vest. You see, it was the custom for some of our crew chiefs at this time to take the back armored plate out and sit on it during flight.

Soon the three of us made it to the beach, out of sight from the shooters. With Saint hard on his heals the bad guy chose to swim for it, and I saw him wade into the South China Sea and disappear under the waves. Saint jumped in immediately after and sank like a rock. This time the flack vest had done him in. I waded out waist deep, and saw an occasional arm and leg fly up from under the bubbling foam. I reached down with both hands and pulled up both a collar and a head of hair. The collar belonged to Saint, and he took a long deep breath. The head of hair belonged to the bad guy. I began pulling them both to the shore when the bad guy tried to take off, and we ended up wrestling on sand. Eventually I was sitting on his chest strangling him, when finally he stopped resisting and relaxed. I remember thinking, "what was I supposed to do now?" Fear set in. I began tearing off his shirt looking for weapons. I was sure he had hand grenades strapped to his chest ready to go off. Didn't find any, just his wound. About this time the squad of Marines showed up. A couple of them took custody of the bad guy, and a sergeant looked at us as if we were two drowned rats that came back from the dead. He explained, "You guys are lucky you weren't killed. Some of our M16's jammed." This was during the time period when M16's were misfiring in the field. A Huey from VMO-6 landed nearby, loaded up the bad guy, and flew off. What were *they* doing here? This was our territory, even if we are the junior squadron. How come we didn't get to take the bad guy in? Shortly afterwards our bird landed. Saint and I climbed aboard, strapped ourselves in, and left the squad of Marines crouching in the rotor wash.

We were still soaking wet. Saint and I looked at each other with an odd stare. We had just done something strange and goofy, and it was just beginning to sink in. It was now that Saint noticed blood on the bulkhead behind me. He told me I was bleeding. I looked over my shoulder and saw the smear. Blood was also running down my arm into my glove. "The S.O.B.s shot me!" I said to the Major. The Huey wobbled when the officers turned around to see what the hell was going on. I hadn't made myself clear - Saint didn't shoot me. When we got back to the base my replacement was waiting and I went to "Delta Med" and got sewed up.

The above event is pretty much the truth as I remembered it, except for those parts that are not. Today it's a story that has grown in the telling by some. I haven't the heart to change their minds. To Saint and all of those who took part, please feel free to make an addendum.

By the way, the wound was a three-inch gash above my right elbow. What I found interesting was not the nick in the arm, but how small the hole was in the flight suit. It was smaller than a pea. I never understood the damage relationship. Also, I had borrowed my buddy Zack's flight suit that day. He complained about that in his book *Farewell Darkness*, and accused me of not washing it afterwards. I wrote to him recently and said, "I don't remember washing *anything* during that year."

# THE GENERAL

### Colonel David H. Hackworth

I've written hard-nosed critiques of Army and Air Force brass who failed in their duty, but here's a story of a different kind of leader - General Edward A. Craig, U.S. Marine Corps, who lived by the principles of leadership and whose first thought was always his warriors.

Retired Marine Norman Kingsley says, "Eddie Craig commanded a Marine unit, Army General Walton Walker's 'Fire Brigade,' during the first months of the Korean War. The brigade was always in the thick of it. Wherever the Reds broke through the U.S. lines, Craig's firemen stopped them cold with hard-hitting counterattacks."

Craig led from the forward foxholes. He ate with rifle squads, and asked the grunts what could be done better. At night he rolled up in a poncho and slept on the ground alongside his Marines. He lived exactly as they did. He felt their pain, saw their problems and looked at the war through their eyes.

During the battles along the Naktong River Craig was always up front, directing the fight and helping the wounded. At the aid stations he went from man to man, holding hands, stroking faces, telling them they'd be okay, and to 'just hang on.' All the time tears streamed down his cheeks. He never tried to hide the tears. That's why his men loved him, fought for him, died for him.

Later, when the Chinese entered the war and surrounded the 1st Marine Division at the Chosin Reservoir, Craig's boss, General Oliver Smith, asked, 'How can we fight against six hundred thousand Chinese?' Craig replied, 'I've been up with the 7th Marines - they're down to one thousand effectives. The 5th Marines have maybe thirteen hundred able to fight. Puller's 1st Marines are still strong, with over two thousand...'

General Smith said, 'We've had it, Eddie.'

Around midnight, with the situation becoming even grimmer and the temperature at twenty degrees below zero, they heard what sounded like a chorus. A recon revealed a squad of nearly frozen Marines singing the Marines' Hymn. Smith said, 'With guys like that, we'll make it, Eddie,' and of course the 1st Division did, adding a brilliant battle honor to the Corps' proud reputation.

Caring leaders like Smith and Craig inspired their warriors to do the impossible. The leadership qualities Eddie Craig displayed didn't jump out of a book. He learned how to lead from the example of his leaders. The pre-WWII generation of Marine skippers, from shavetail to general, wore the same dirty green dungarees, carried their own packs and bedrolls, and lived by the rule, "Know and care for your men, and always set the example."

After he retired, Craig still followed this axiom. Looking at a photo album, he'd recall the names of his Marines as if it were yesterday. "That's PFC Will Hurstman, here's Corporal. Brandenhorst and there's Sal Di Carlo - baked the best bread I ever ate." Those pictures dated back sixty years to campaigns in Haiti, Santo Domingo, Nicaragua, China, Philippines and WWII.

Until his death last December, the General kept in touch with hundreds of Marines who served with him. A few years ago one of his sergeants, who had been crippled during the battle of Seoul, was having a hard time putting a daughter through college. Craig moved a few mountains and got her a scholarship.

Since the Vietnam War, most U.S. generals have become carbon copies of corporate managers. They're too busy being executives to spend much time down in the mud with their warriors. In eight years of combat in Vietnam, there is not one case of a U.S. general officer spending a night with a rifle company. During the Gulf War, Somalia and Haiti, the only generals I saw down at the grunt level were the "streakers" - a chopper lands, then a general trots around the perimeter saluting, patting, shaking and chewing - ten minutes, in and out.

It's no damn wonder today's top is not in touch with the bottom. The brass must get back to the basics. Eddie Craig would be a good role model, and there are others, from "Stonewall" Jackson to Hank "The Gunfighter" Emerson, including a dozen or so serving generals - but not enough to provide our warriors with the Craig-like leadership they deserve.

**This article first appeared on February 28, 1995. Used with permission of Colonel David H. Hackworth.**

# SHORT-TIMER
## *The Killing of Gus Hasford*

Grover Lewis

"The best work of fiction about the Vietnam War," *Newsweek* called Gus Hasford's *The Short-Timers* when it was first published in 1979. The slim hardcover sold, like most first novels, in the low thousands, but established its author as one of the premier writing talents of his generation. In the tradition of Stephen Crane, Ernest Hemingway and James Jones, the book summoned up the horrors of war in an unrelenting voice with all the potential for world-class success.

Hasford's critical stock rose even higher when Stanley Kubrick filmed the book as *Full Metal Jacket.* Released in 1987, the picture received one major Academy Award nomination - shared by Kubrick, Michael Herr and Hasford himself - for best screen adaptation. At a stroke, the struggling, rootless young novelist entered the upper realms of "A-list" Hollywood, but in a skein of envy, spite and the inexorable grinding of bureaucratic "justice" - all of them compounded by Hasford's own obsessive passion for books - his newfound celebrity backfired and he was sent to jail on bizarrely exaggerated charges involving stolen and overdue library books. It all combined to kill him.

Gus died alone, as he had mostly lived, in Greece on

January 29, 1993 at the age of forty-five from the complications of untreated diabetes. His death coincided eerily with the twenty-fifth anniversary of the Tet offensive, the campaign so graphically described in *The Short-Timers*. Two weeks after the shock of his death, twenty-odd mourners gathered in the chapel at Tacoma's Mountain View Memorial Park. Gus' kin sat close to the front - his mother Hazel, a gaunt and visibly ailing Alabama native, along with Gus' younger brother, Army Sergeant Terry Hasford, and Terry's Korean wife, Soo. In back of them a couple of rows were the Snuffies, a cadre of Gus' brothers-in-arms from the Vietnam days, all wearing their battle ribbons on sweaters or lapels, with the five men who'd managed to attend representing a total of eight Purple Hearts.

Several editions of *The Short-Timers* lay on Gus' bier, along with copies of his later books, *The Phantom Blooper* and *A Gypsy Good Time*. Gus' picture as a stern-faced teenage Marine sat on a pedestal, the urn containing his ashes on another. His Vietnam decorations were also on display - two rows of ribbons plus-one, the highest being the Navy Achievement Medal with a Combat "V."

Five wreaths ringed the dais - two from Gus' family, another from "Doctor Dave" Walker, who'd been Gus' landlord and unofficial medical adviser during his last days in America. A spray of white tulips was signed, "From all the gang down at the Cafe Cafard." The floral tribute from the 1st Marine Division ISO Snuffies spelled out SEMPER GUS. Nothing from Hollywood - not a bud or sprig from Stanley or Michael or any of the other distant A-listers who'd profited from *The Short-Timers*.

Assorted friends sat across the aisle - Doctor Dave, and book dealer Bruce Miller from San Luis Obispo who'd supported Gus during his trial, and Kent Anderson, another formidable Vietnam War novelist. Anderson, author of *Sympathy for the Devil*, fidgeted in agitation. I sat just behind him, shivering a little in my lightweight L.A. clothes. In the far rear of the room, with their rifles stacked out of sight, sat six young Marines in full dress uniform, white covers squared on their blue-clad knees.

The noon ceremony was spare, simple, and elegantly offbeat. Steve "Bernie" Bernston, chief archivist of the Snuffies, spoke a brief eulogy and then set out bottles of Jack Daniels, fruit juice, Evian water and California wine. Nine other mourners, including myself, offered personal tributes to Gus, concluding with toasts to his memory. A local Presbyterian minister, a little nonplussed by the procedure, toasted God.

At the service's conclusion the Marine honor guard fired four volleys of salute outside the chapel, followed by a bugler playing taps. A smart-stepping Marine SNCO presented Gus' mother with a folded American flag. "On behalf of a grateful nation, ma'am, we present this flag as a token of your son's honorable and faithful service to the United States of America." Mrs. Hasford sat with her eyes lowered, softly fingering the cloth. "I never could understand that boy," she'd told one of the Snuffies a few days before, "just never could."

In a caravan of cars, the memorial moved en masse to

Bernston's house in a nearby suburb, where the post-mortems continued through the afternoon and into the evening in a glow of sipping whiskey, fond remembrance and brusque camaraderie. Many of the characters in *The Short-Timers* had been modeled on the now-middle-aged Snuffles, and the men were strapping proud of the distinction. In Vietnam with Gus, they'd all been Marine combat correspondents, equally adept at filing dispatches or fighting hooch-to-hooch. At Gus' wake, circulating from bar to buffet, they openly discussed his jail sentence and its effect on him. None of them approved of his transgressions, but none of them had rejected him, either. As men who'd shared life at its worst, they viewed Gus as family - and whatever had happened, they loved him. "Capital punishment... for library violations?" Gordon Fowler growled. Gordon was "Cowboy" in the book.

Bernie told perhaps the best "Gus" story. "It's peculiar, but this happened exactly twenty-five years ago today. I'd set up a base camp in Hue City, and Walter Cronkite rolls up with a camera crew. He was doing a stand-upper with some pogue colonel, asking about rumors that our guys had been looting. Just then Gus busts in with two black onyx panthers and a stone Buddha on his back. 'Hey, there's a whole temple full of this shit!' he hollers. 'We can get beaucoup bucks for this stuff in Saigon!' I hustled him outside quick, and Cronkite of course came back home and declared the war unwinnable on national TV."

There were a million Gus stories, and some of the classic ones were told by Major Mawk Arnold, USMC (Ret). He arrived late in the afternoon, delayed in transit, but was essential not only for his moral presence, but to carry Gus' remains back to the Hasford family plot in Haleyville, Alabama. Picture John Huston with a shiny pate, resplendent in dress blues with ribbons dating back to the South Pacific and China in the 1940s. "Skipper" Arnold had created the Snuffy team by letting them make their own mistakes and victories. Nearing seventy now, he'd been the catalyst - the force who'd molded raw young recruits with reading habits and verbal skills into warrior artists in a World of Shit.

Another round of toasts commenced after dinner. Every Snuffy present had helped Gus out of various hapless jams during, and ever since, the war... and Jesus, if the SOB hadn't slipped out of their reach, maybe he wouldn't have died. It hung over the table, unspoken. Bob Bayer ("Mr. Short-Round") recalled driving hundreds of miles to rescue Gus from his latest broken-down lemon car. "He could start out to meet you with a thousand bucks in his pocket, walk past a bookstore, and then you'd have to spring for dinner." Earl Gerheim ("Crazy Earl") nodded and smiled. "Gus had a forty-five-year childhood - the childhood the rest of us missed, I guess." There was general agreement that Gus had been a zany, wonderful, generous, naive, impractical, homemade genius, maybe too pure in his ways to die of old age. Bernie raised his wine glass. "To those of us who are near," he said, "and those far away, and those who are beyond the wire."

This story originally appeared in *LA Weekly* on June 27, 1993.

# UNSUNG FILMMAKER
## *Of Iwo Jima*

Hal Buell

*A few years ago I heard they had found the cave containing Bill Genaust's remains, but decided it was best to leave it undisturbed. I also had an amazing experience looking at the famed flag raising photo, courtesy of Bill Genaust and Joe Rosenthal. Someone found and printed the movie frame which coincided precisely with the Rosenthal photo, and if you placed the two side-by-side and looked at them cross-eyed, the photo became three dimensional - since they were taken at the same moment from slightly different angles.*

Sixty years ago this week a U.S. Marine motion-picture cameraman stood in dust and bramble inside a rocky volcano, waiting to film a flag-raising. "I'm not in your way, am I, Joe?" he shouted to a nearby still photographer over the relentless Pacific wind.

"No, it's all right," the photographer replied. "Hey, there she goes, Bill!"

Five Marines and a Navy corpsman pushed up the long, heavy pipe which had been improvised as a flagstaff. The wind snapped the flag during its rise, but once up Old Glory stood out straight and full.

The movie man cranked one hundred and ninety-eight

frames of 16mm Kodachrome ASA 8 film through his Bell & Howell camera until the film ran out, but he would never know whether he captured the entire lift. The still photographer took a single picture with his 4x5 Speed Graphic at the peak of action.

Each of the two photographers had caught an enduring moment of the American experience. Joe Rosenthal, the *Associated Press* still photographer, would win a Pulitzer Prize for his shot of the raising of the flag on the summit of Iwo Jima's Mount Suribachi on Feb. 23, 1945. The film by Sergeant Bill Genaust would live on gloriously as well, but his name would be all but lost to history after his death nine days later.

Genaust's film sequence, which he did not live to see, was widely shown in movie houses and later on television, but for decades he was not officially recognized as the cameraman who shot the famous footage. Unknown by most and forgotten by many, Genaust was - and remains today - a Marine left behind on that distant island.

Iwo Jima is not a pretty place. Its craggy topography is dominated by the extinct volcano, Mount Suribachi, a scarred hump that rises 556 feet above sea level. It lacks the soaring grace of Japan's Mount Fuji, or the majesty of our own Rockies.

On the morning of February 19, 1945, Genaust and Rosenthal rode toward the island with the Marines but in separate landing craft. Suribachi's Japanese gun installations were trained on the black, volcanic sands of the beach and created a hell fire. Rosenthal would later say, "Survival was like walking in rain without getting wet," but the two men dodged the bullets and despite casualties comparable to Normandy, survived the assault.

On the fifth day of the battle, Genaust met Marine still photographer Bob Campbell and Rosenthal at the base of Suribachi. They had all heard a flag would be raised on the summit, and wanted to photograph this key taking of the island's high ground.

Halfway up the mountain, they met *Leatherneck* magazine photographer Sergeant Lou Lowery coming down. "You're late," Lowery said. "The flag is already up." The three men then believed they would not get photos of a flag going up, but hoped that another picture would be possible. Once at the top, the trio found Marines preparing a second flag - a larger flag, a flag they said "could be seen by every Marine on the island."

Genaust and his companions positioned themselves for the picture. Rosenthal placed himself head-on, and Genaust stood about an arm's length from Rosenthal's right side and slightly forward. Their pictures were flown to Guam, and Rosenthal's photo - transmitted to the world - was an immediate sensation.

It took weeks to process Genaust's Kodachrome which, once released, was equally successful in lifting the spirits of a war-weary home front eager for victory and impatient with rising casualties from the Pacific. The film was shown in movie houses and appeared daily for years as an overnight signoff segment on TV stations - but there was no official recognition of the photographer. A quirk in regulations authorized bylines for still photographers, but decreed that film would be distributed without a photographer's credit.

Rosenthal saw his picture for the first time on March 4 on Guam, where he had been sent by the *Associated Press*. That same day back on Iwo Jima, a B-29 bomber - battered during a bombing run over Tokyo - made an emergency landing. Normally Genaust would have photographed the bomber, but the weather was poor - overcast, dark and misty - so instead he was on Hill 362A, where Marines were mopping up any remaining resistance. With photography impossible, Genaust turned to his carbine and .45 pistol to help his buddies. Genaust and another Marine ducked into a cave to escape the heavy rain, and when he turned on a flashlight to check his surroundings the Japanese hidden in the cave opened fire, killing the two Marines instantly. Other Marines cleaned out the cave with flamethrowers, and bulldozers blocked up the entrance.

For decades, Genaust remained anonymous. Efforts by friends and colleagues, urging the Corps to see past the regulations, were to no avail. Finally, forty years later, Marine brass issued a letter of appreciation for exemplary camerawork and heroism and officially recognized the photographer. Genaust's friends prepared a plaque, and in 1995 it was installed atop Mount Suribachi. Bill Genaust finally took his rightful place alongside Joe Rosenthal as the men whose pictures immortalized the bloodiest battle in Marine Corps history.

Marine casualties at Iwo Jima included nearly six thousand dead and about eighteen thousand wounded. More than twenty-one thousand Japanese were killed or committed suicide. Of the twelve Marines who raised the two flags on Suribachi, six later died in the battle and four were wounded.

True to the Corps' tradition of recovering their dead, most of the Marines killed and initially buried on Iwo Jima were returned to the U.S. by the 1950s. However, the cave where Genaust died was considered too dangerous to open because of possible explosives, and its entrance eventually was lost to time.

The island was returned to the Japanese in 1968, and today Old Glory flies only four days each year - but every time the film of the flag-raising appears, as it does occasionally in documentaries, viewers now will know of Sergeant Genaust. He was one Marine who immortalized on film his nation's fight for freedom, and his Corps' honor - though he remains behind, entombed forever on Hill 362A in a forgotten cave without a marker.

**Published February 20, 2005. Hal Buell is a writer, lecturer and award-winning photo editor who was head of the *Associated Press* photo services for more than twenty years.**

# 1/400<sup>th</sup> OF A SECOND

Mitchell Landsberg

Fifty years ago this month, a young *Associated Press* photographer named Joe Rosenthal shot the most memorable photograph of World War II - a simple, stirring image of five Marines and one Navy corpsman raising the flag at Iwo Jima. It took but a sliver of time: 1-400th of a second. Yet it has consumed the past half-century of Joe Rosenthal's life.

He has been called a genius, a fraud, a hero, a phony. He has been labeled and relabeled, adored and abused, forced to live and relive, explain and defend that day atop Mount Suribachi on each and every day that has followed, more than eighteen thousand and counting. "I don't think it is in me to do much more of this sort of thing," he said during an interview - his umpteen-thousandth - about Iwo Jima. "I don't know how to get across to anybody what fifty years of constant repetition means."

Rosenthal is eighty-three now, nearly blind, a pudgy man with a dapper white mustache and a horseshoe of white hair curving around the back of a largely bald head. He lives alone in San Francisco near Golden Gate Park, in a little apartment largely given over to stacks of correspondence and documentation related to Iwo Jima.

In 1945 he was thirty-three, too nearsighted for military service, short and athletic, with a brushy brown mustache and a head full of tight brown curls. As an AP photographer

151

assigned to the Pacific theater of the war, he had already distinguished himself - and shown a streak of bravado - in battles at New Guinea, Hollandia, Guam, Peleliu and Angaur. No one remembers Rosenthal for those pictures now. There is only Iwo. Bloody Iwo. It is the battle that Joe Rosenthal will fight until he dies.

We remember Iwo Jima for two good reasons. One is that it was the costliest battle in Marine Corps history. Its toll of 6,821 Americans dead, 5,931 of them Marines, accounted for nearly one-third of all Marine Corps losses in all of World War II. The other is Joe Rosenthal's picture.

It has been called the greatest photograph of all time. It may well be the most widely reproduced. It served as the symbol for the Seventh War Loan Drive, for which it was plastered on 3.5 million posters. It was used on a postage stamp and on the cover of countless magazines and newspapers. It served as the model for the Marine Corps War Memorial in Arlington, Virginia, a symbol forever of the valor and sacrifices of the U.S. Marines.

As a photograph, it derives its power from a simple, dynamic composition, a sense of momentum and the kinetic energy of six men straining toward a common goal, which for one man has slipped just out of grasp. "It has every element... it has everything," marveled Eddie Adams, a former AP photographer who took another picture that helped sum up a war - one of a South Vietnamese police chief executing a suspect. Of Rosenthal's picture, he added, "It's perfect. The position, the body language... you couldn't set anything up like this - it's just so perfect."

And therein lies the problem. Some people think Rosenthal's picture is *too* perfect. For fifty years now, Rosenthal has battled a perception that he somehow staged the flag-raising picture, or covered up the fact that it was actually not the first flag-raising at Iwo Jima.

All the available evidence backs up Rosenthal. The man responsible for spreading the story that the picture was staged, the late *Time-Life* correspondent Robert Sherrod, long ago admitted he was wrong - but still the rumor persists.

In 1991, a *New York Times* book reviewer, misquoting a murky treatise on the flag-raising called "Iwo Jima: Monuments, Memories and the American Hero," went so far as to suggest that the Pulitzer Prize committee consider revoking Rosenthal's 1945 award for photography.

And just a year ago, columnist Jack Anderson promised readers "the real story" of the Iwo Jima photo - that Rosenthal had "accompanied a handpicked group of men for a staged flag raising hours after the original event." Anderson later retracted his story, but the damage, once again, had been done.

Rosenthal's story, told again and again with virtually no variation over the years, is that on February 23, 1945, four days after D-Day at Iwo Jima, he was making his daily trek to the island on a Marine landing craft when he heard that a flag was being raised atop Mount Suribachi, a volcano at the southern tip of the island.

Marines had been battling for the high ground of Suribachi since their initial landing on Iwo Jima and now, after suffering terrible losses on the beaches below it, they appeared to be taking it.

Upon landing, Rosenthal hurried toward Suribachi, lugging along his bulky Speed Graphic camera, the standard for press photographers at the time. Along the way he came across two Marine photographers, PFC Bob Campbell shooting still pictures, and Staff Sergeant Bill Genaust shooting movies. The three men proceeded up the mountain together.

About halfway up, they met four Marines coming down. Among them was Sergeant Lou Lowery, a photographer for *Leatherneck* magazine, who said the flag had already been raised on the summit. He added that it was worth the climb anyway for the view, so Rosenthal and the others decided to continue.

The first flag, he would later learn, had been raised at 10:37 AM, and shortly thereafter Marine commanders decided - for reasons still clouded in controversy - to replace it with a larger flag.

At the top, Rosenthal tried to find the Marines who had raised the first flag, figuring he could get a group picture of them beside it. When no one seemed willing or able to tell him where they were, he turned his attention to a group of Marines preparing the second flag to be raised.

Here, with the rest of the story, is Rosenthal writing in *Collier's* magazine in 1955. "I thought of trying to get a shot of the two flags, one coming down and the other going up, but although this turned out to be a picture Bob Campbell got, I couldn't line it up. Then I decided to get just the one flag going up, and I backed off about thirty-five feet. Here the ground sloped down toward the center of the volcanic crater, and I found that the ground line was in my way. I put my Speed Graphic down and quickly piled up some stones and a Jap sandbag to raise me about two feet (I am only five feet, five inches tall) and I picked up the camera and climbed up on the pile. I decided on a lens setting between f-8 and f-11, and set the speed at 1-400th of a second.

At this point, 1st Lieutenant Harold G. Shrier... stepped between me and the men getting ready to raise the flag. When he moved away, Genaust came across in front of me with his movie camera and then took a position about three feet to my right. 'I'm not in your way, Joe?' he called.

'No,' I shouted, 'and there it goes.'

Out of the corner of my eye, as I had turned toward Genaust, I had seen the men start the flag up. I swung my camera, and shot the scene."

Rosenthal didn't know what he had taken. He certainly had no inkling he had just taken the best photograph of his career. To make sure he had something worth printing, he gathered all the Marines on the summit together for a jubilant shot under the flag that became known as his "gung-ho" picture.

And then he went down the mountain. At the bottom, he looked at his watch. It was 1:05 PM. Rosenthal hurried back to the command ship, where he wrote captions for all the pictures he had sent that day, and shipped the film off to the military press center in Guam. There it was processed, edited and sent by radio transmission to the mainland.

On the caption Rosenthal had written, "Atop 550-foot

Suribachi Yama, the volcano at the southwest tip of Iwo Jima, Marines of the Second Battalion, 28th Regiment, Fifth Division, hoist the Stars and Stripes, signaling the capture of this key position." At the same time he told an AP correspondent, Hamilton Feron, that he had shot the second of two flag raisings that day and Feron wrote a story mentioning the two flags.

The flag-raising picture was an immediate sensation back in the States. It arrived in time to be on the front pages of Sunday newspapers across the country on February 25. Rosenthal was quickly wired a congratulatory note from AP headquarters in New York - but he had no idea which picture they were congratulating him for.

A few days later, back in Guam, someone asked him if he posed the picture. Assuming this was a reference to the "gung-ho shot," he said, "Sure." Not long after Sherrod, the *Time-Life* correspondent, sent a cable to his editors in New York reporting that Rosenthal had staged the flag-raising photo. *Time* magazine's radio show, "Time Views the News," broadcast a report charging that "Rosenthal climbed Suribachi after the flag had already been planted... like most photographers (he) could not resist reposing his characters in historic fashion." *Time* was to retract the story within days and issue an apology to Rosenthal. He accepted it, but was never able to entirely shake the taint *Time* had cast on his story.

A new book, *Shadow of Suribachi: Raising the Flags on Iwo Jima*, offers the fullest defense yet of Rosenthal and his picture. In it, Sherrod is quoted as saying he'd been told the erroneous story of the restaging by Lowery, the Marine photographer who captured the first flag raising. "It was Lowery who led me into the error on the Rosenthal photo," Sherrod told the authors, Parker Albee Jr. and Keller Freeman. "I should have been more careful."

Rosenthal, who was to become close friends with Lowery in the years after Iwo Jima, rejects this explanation. "I think that is a twist that has been put on by Sherrod," Rosenthal said. He believes the source of the misunderstanding was his response to the question about his picture being posed.

It is probably moot. Rosenthal is the only party to the dispute who is still alive. His attitude now is mostly one of forgiveness and acceptance. So many years, after all, have passed. There is still, of course, the issue of whether the second flag-raising was noteworthy enough to have been enshrined as a historical icon. Here, the facts are of little use - all that matters is interpretation. To be sure, it didn't help that the Marine Corps and most of the wartime press conveniently glossed over the fact of the first flag-raising. This helped foster a public notion of cover-up - but whether or not there was a cover-up (Albee and Freeman are persuasive in arguing that the Marine brass decided to put a lid on the first flag-raising), was the second flag-raising worthy of Rosenthal's picture? Some vehemently argue no.

"They call that the Iwo Jima flag-raising, which it ain't," declared Charles Lindberg, a retired electrician in Richfield, Minnesota, who is the last surviving member of either flag-raising - in his case, the first. "It's a good picture," Lindberg conceded. "I even told Joe Rosenthal that it was a good picture. But me and him get into a few arguments." That is because Lindberg, like others in the first-flag raising, believed that all the glory was showered on the second flag-raisers, who were less deserving.

Rosenthal doesn't argue that one group was more deserving than another. "In my own opinion, any one of those troops who had their feet on Iwo Jima is a hero."

The fact is that neither set of flag-raisers encountered serious resistance from the Japanese as they scaled Mount Suribachi that day. And in retrospect, the scaling of Mount Suribachi was not the great turning point in the battle that it may have seemed - the fighting on Iwo Jima continued for thirty-one more days.

The other undeniable fact is that both groups took part in some of the fiercest combat of the war. Five of the eleven men in the two flag-raisings never left Iwo Jima alive.

Perhaps the best argument for Rosenthal's photo is simply that it is powerful on a symbolic, not a literal, level. Americans responded to it because it was a stirring image of the victory they so badly craved. On that level, it is unassailable.

Marianne Fulton, chief curator of the International Center of Photography at George Eastman House in Rochester, New York, said the photo must be seen in the context of a perilous time. "You're worried about your life, your family, the future of the nation, and this really incredible picture of strength and determination comes out. A picture like that is a real gift."

For Joe Rosenthal, Iwo Jima brought fame but not

fortune, acclaim but not overwhelming success. He spent the rest of his career as a workaday photographer at the *San Francisco Chronicle,* shooting politicians and drug dealers, fires and parades. He scoffs at the notion of being included among the "great" photographers - but they include him in their company. Carl Mydans, the renowned *Life* magazine photographer, offers this explanation for Rosenthal's immortality. "If you can get the right moment, the instant, it stays around forever."

And so it will be with Rosenthal. He doesn't have a copy of the Iwo Jima picture hanging in his apartment, only an etching of it and two cartoons lampooning it. He is modest to the point of self-deprecation. Still, when he was once asked if he would rather that some other photographer had taken the flag-raising shot, he shot back, "Hell, no! Because it of course makes me feel as though I've done something worthwhile. My kids think so – that's worthwhile."

On one wall of Rosenthal's cluttered living room is a framed photograph of seven bleary war correspondents at Guadalcanal. They have just stumbled out into the bright light of morning after a night of drinking and card-playing. If they felt the way they look, it had been a long, long night. In the center of the picture is Rosenthal, scratching the stubble on his chin, looking a little bemused and a little cockeyed, while a *Newsweek* correspondent next to him drapes a hand over his shoulder for support. The whole picture has a washed-out, overexposed look, perfectly matching the mood of its squinting subjects.

Standing now in his living room, Rosenthal looks at it

fondly. "That," he declares with a proud chortle, "is the greatest photograph of World War II."

**On April 13, 1996 Rosenthal was named an honorary Marine by then Commandant of the Marine Corps General Charles C. Krulak, on August 20, 2006, at age ninety-four, Rosenthal died of natural causes in his sleep at a center for assisted living in Novato, a suburb of San Francisco, and on September 15, 2006 he was posthumously awarded the Distinguished Public Service Medal by the Marine Corps.**

# A MARINE'S STORY

Ron Zaczek

In 1984, I was halfway through seven years of therapy for Post Traumatic Stress Disorder. My Vet Center counselor and I faced a problem. For over three years we'd been unraveling the missions, mortar attacks, and nights on the perimeter and the unhappy days of homecoming that were the building blocks of PTSD. With a lot of help, I'd done a fair job remembering the things which had changed my life seventeen years earlier. I had corresponded frequently from Vietnam, averaging more than a letter each day, even if the destination and return addresses on the envelope were sometimes longer than what I'd scrawled inside. Friends had saved the letters, and I'd been using them to reconstruct memories. PTSD isn't about having thirteen months worth of bad hair days in-country. Typically it's three or four especially bad events, some lasting only a few long moments, some hardly more than an hour. During counseling, the challenge is for a vet and his counselor to discover exactly which three or four moments out of hundreds were so extraordinary that they could influence behavior in seemingly unconnected ways so many years after the war. Memories, and often knowing the date and sequence in which events occurred, are crucial to coming to terms with the past.

I'd discovered a two-week lapse in my letters home. No

letters to anyone, for almost fourteen days, from an inveterate writer. And the letters that framed the lapse seemed to be written by two different men, with very different feelings about what he was doing in the war. What had happened? Weeks of discussion ended with no progress. Indeed, three years of steady progress seemed about to be derailed. My therapist had an idea, and called Headquarters, Marine Corps. "I'm working with a Marine in trouble. Do the Marines have records, historical information - anything, that can help this Marine to remember?"

The officer at HQMC answered crisply. "He's a Marine. We take care of our own."

The next day the Chief Historian of the Marine Corps phoned my counselor and placed the services of the USMC Historical Center at our disposal. The staff of the Center in the Washington Navy Yard treated us royally. Examining Command Chronologies, After Action Reports and Oral History tapes, I was able to uncover the memories hidden in those lost weeks. Dealing with them was painful, but therapy continued. More than anything though, I was proud of once again touching the spirit that made us all Marines.

I was no lifer, and finished my four years as possibly the only E-5 NATOPS crew-chief-instructor and engine man who ever devoted one hundred percent of his bad attitude to running a coffee mess and loan shark operation on the flight line at New River. I unhappily closed out those years in the "Crotch" as, well, pretty much a screw-up - and yet seventeen years after my war, the Marines still took care of this very least of their own. Semper Fidelis, Marines.

# NAME THAT HERO

**Donna Miles**

*This has long been one of my pet peeves. If I were king the Oscars - which are nothing more than awards given to narcissists for making believe they are someone else and being paid millions to do it - would be preempted and replaced with the presentation of Purple Hearts.*

The challenge issued by a flight attendant during a recent commercial air flight seemed innocuous enough - "Name just one of the (then) five Medal of Honor recipients from the current engagements in Afghanistan or Iraq, and get a free drink coupon" - but the passengers response... more specifically the inability of all but just one to respond... revealed how little the average American knows about our military heroes.

Bombarded by superhero lore almost from birth, many Americans grow to revere fictional heroes as well as sports and celebrity icons - but silence descended over the cabin of a flight bound from Jacksonville, Florida to Baltimore when the conversation turned to those who had earned the nation's highest honor for valor - even when a free cocktail hung in the balance.

Dale Shelton, an Annapolis, Maryland resident who served five years as a Navy intelligence specialist, was the only passenger to press the button over his seat to beckon the attendant. Shelton's response - Army Sergeant First Class Paul R. Smith, the first Medal of Honor recipient in the global war on terror and in Operation Iraqi Freedom.

Smith received the highest military honor for valor posthumously on April 3, 2005, two years to the day after saving more than one hundred soldiers in the battle for Baghdad's airport. His young son and widow accepted the award on his behalf during a solemn White House ceremony.

The flight attendant gave free drink coupons to Shelton, as well as his wife Jean and two other traveling companions. Then he returned to crew area to announce over the intercom that only one person had correctly answered the challenge.

This time, the attendant offered a second challenge - "Name an *American Idol* winner." The cabin lit up like a pinball machine as forty-three passengers scrambled to push their attendant call button. Passengers named various *Idol* winners, and the attendant announced he wasn't going to award drink coupons for that answer, telling them that "naming an *Idol* winner was not worth a free drink."

He concluded his announcement with a question. "What's wrong with our country when out of one hundred and fifty passengers, only one can name a Medal of Honor recipient, but forty-three can name an *American Idol* winner?"

Later during the flight, Shelton shared with the attendant his own frustration over "the current lack of appreciation of our military heroes." The attendant asked Shelton if he knew the names of the other four Medal of Honor receipts from current military operations, and he was able to name three - Navy Lt. Michael Murphy, Navy Petty Officer 2nd Class

Michael Monsoor and Army Spc. Ross McGinness. All were killed sacrificing themselves to protect their comrades during enemy attacks.

Murphy, a Navy SEAL, died trying to save his team members during Operation Red Wing in Afghanistan. Monsoor, also a SEAL, died in Iraq on while using his body to absorb a grenade blast that likely would have killed two nearby SEALs and several Iraqi soldiers. McGinnis died after throwing himself on a hand grenade in Iraq to save four fellow soldiers when insurgents attacked their Humvee.

Shelton said he regretted that he had forgotten the name of Marine Corporal Jason Dunham, who died using his body to shield fellow Marines in Iraq from a hand grenade.

The flight attendant didn't hold Shelton's memory lapse against him. "He gave me all the remaining drink coupons he had in his possession, and shook my hand," he said.

From the *American Forces Press Service*, March 18, 2009. Note: A new special report on the Defense Department home page pays tribute to U.S. servicemembers who have earned the Medal of Honor for action in the war on terror.

# BACK TO BASICS

Captain G. E. Rector

Carl Sandburg once wrote, "When a nation goes down, or a society perishes, one condition may always be found - they forget where they came from." Our nation is not about to go down, and we're not about to perish, but we have lost sight of some of the things that helped make this country what it is today. Certain basic characteristics are necessary if a nation is to be great, and is to remain great. It is on this basis that the U.S. Marine Corps assumes its unique importance. Aside from providing the world's finest military training, the vital importance of the Marine Corps lies in the fact it is one of the few strongholds of old-fashioned virtues in the United States today because every Marine must understand the fundamentals of discipline, honor, valor, and duty.

We take recruits, strip them, scrub them, and shear off their hair. We begin making them Marines who are proud to be citizens of the greatest country in the world. There are just twelve weeks of recruit training, and since we unfortunately have to allow for sleeping and eating, we measure our curriculum in terms of hours and fractions of hours. The recruit learns, for the first time perhaps, that reality is not something to make fun of and rumors to the contrary, God is not dead. Grace is said before each meal.

A Marine recruit learns the meaning of professionalism, standing ready to meet our country's needs at any time and doing so coolly and capably. The recruit is trained neither to hate, nor is he "whipped up" emotionally for battle. Marines are not dangerous to anyone, except to the enemy they face. At a time when it seems we are suffocating from all the "isms" that afflict modern society, a Marine concerns himself with only two - patriotism and professionalism.

Each Marine learns the meaning of discipline - the attitude which ensures prompt obedience to orders, and in the absence of orders, the initiation of appropriate actions. He learns that discipline is a way of life, and that his own life may depend upon it.

When the Fifth and Seventh Regiments of Marines were fighting their way down from the Chosin Reservoir in Korea, a blown-out bridge made Sinchilin Pass impassable. The yawning chasm had to be crossed if those seven thousand Marines - many of them wounded and frostbitten - were to continue down to safety. The Marine Engineers air-dropped bridging, and the infantrymen came down along with their dead and wounded. They did it with the moral discipline which is the Marines' antidote to fear and despair. When aching muscles and bursting lungs screamed "Quit!" their hearts and their guts shouted back "Do it!" - and do it they did, as a fighting team.

A Marine learns other things, too. He learns that a man's word is his bond. He learns that success is not how much you can get away with, and that you don't cheat, steal, and lie. Most of all you are a Marine, and proud of it.

Intellectual "jet-setters" may consider us to be hopelessly archaic (if they're even aware of our existence). Actually, we're remarkably uncomplicated people. We believe in things, and tend to be a bit sentimental and rather simple - the type for whom morning and evening colors are important rituals. We are the type who snap to attention at the opening notes of the *Star-Spangled Banner,* and as long as we live we will stand when they play *The Marines' Hymn.* We remember "things endured and things achieved, such as regiments hand down forever."

The next time someone looks at me and says, "Oh, you're a Marine," I'll thank him and proudly say, "Yes, I am!" Our men don't have to go looking for employment. There are always jobs for men of their caliber - but the price they pay for the privilege of serving their country can be high. There are American graves all over the face of the Earth, and our Marines fill their share of those graves. They never asked what their country could do for them, and now they never will.

Yes. I'm a Marine - and damn proud of it!

**Captain Rector's letter appeared in *Proceedings* in November 1983.**

# JARHEAD

*A lot of people saw the movie 'Jarhead' and think they know what the Corps is about, but like most third-hand accounts it didn't paint a very accurate picture of reality. It seems to me Anthony Swofford sold out the Marine Corps for a truckload of cash, because as we all know sensationalism is what sells in Hollywood. It has been a while since this movie came out, but the concept of a Marine turning on his own for money still sticks in my craw. But don't take my word for it - here are a few typical "reviews" from around the Corps:*

When I first saw that Anthony Swofford's travesty of a book *Jarhead* was being made into a movie, I was sickened. His book was bad enough. I remember reading it, and as a Marine myself who *also* served in Desert Storm, I thought in the back of my mind that it probably wouldn't be long until the Hollywood left came to this book to make it a movie. Also, Swofford is the type of bottom-feeding traitor who would do just about anything for a buck, so it's no surprise there either. But, I digress.

This movie/book is an abomination. Point blank and period. It's not "realistic" in any sense, other than perhaps the color of desert fatigues. You want a real sense of life at USMCRD? Go get a copy of *The D.I.* with Jack Webb. Or at least watch nothing past the first thirty minutes of *Full Metal Jacket*. Anything past that is garbage too. That's how it is, and how it was for me at P.I. I can't say for my Hollywood Marine brothers in San Diego. You want to *really* get an idea what it's like? Enlist. Anything else and you're really selling the experience short, take my word.

Desert Storm itself was nothing like this. And coming from a Marine who still has brothers fighting in Iraq, I take this movie/book as a personal attack against all Marines. It truly is, and always has been since Swofford wrote it. Do yourself a favor and avoid it all costs.

If you've seen this movie and you're inspired to read something, I urge you *not* to read Swofford's book. Instead go pick up a copy of *Marine* and read about Chesty Puller. Then watch *The Sands of Iwo Jima* and thank God that you live in the USA and still have sensible brothers and sisters in the Armed Services (not just Marines, though I am biased after serving in the USMC 12+ years), who hold this country's values near and dear.

In 1987 *Time* magazine ran an infamous cover that consisted of a Marine in his dress blue uniform with a blackened eye. The cover was intended to depict the shame befallen the Marines after the Clayton Lonetree spy scandal and it was met with outrage. How dare *Time* sucker-punch the entire Marine Corps because of the crimes of just one of its members? Yet after seeing the movie *Jarhead*, Anthony Swofford's autobiographical account of the Marines during the first Gulf War, a black eye is the least of the Marines'

problems.

The fundamental theme of *Jarhead's* portrayal of Marine life is that heroes do not exist. One cannot depict the Marine Corps accurately without noting that at least some of its members perform feats of strength, endurance and bravery, and that to build an entire institution of such men, certain virtues are required. Yet like Stanley Kubrick's *Full Metal Jacket*, a movie acclaimed for its supposed depiction of Viet Nam-era Marines, none of these men and certainly none of these virtues are to be found.

Instead, what one finds in *Jarhead* are empty men who drift though life denied of what they truly want, and who choose to make up for it with emotional outbursts and sadistic and debased pleasures. Again and again, this is what Hollywood sees when it looks at the Marines.

I served five years in the Marine Corps during the time *Jarhead* was set and I can certainly recount stories, both humorous and horrific, but overall if I had to characterize my and my fellow Marines' service it would be honorable commitment to the betterment of one's self and the defense of the American nation. The men I worked with might not have talked about it every day. There might have been the occasional breach of conduct or character, and some may have even failed miserably in achieving the standard of excellence that is the hallmark of the Corps, yet overall (and in the metaphysically significant sense - the only sense that matters in art) almost every Marine I knew was in the Corps for a purpose and that purpose was good, noble, and just.

That's why I for one was proud to wear the Marine

uniform, and that's what no Hollywood movie that I know of has ever been able to accurately capture on film. Given the freedoms the Marines have fought so valiantly over their history to preserve, it's a tragedy they haven't received better treatment from Hollywood in return.

After viewing this horrid piece of trash, I'd agree with the following from a wonderful soldier, Sergeant Jeff Davids, 1983-1987:

The movie *Jarhead* is a disgraceful attack on the dignity and honor of all current and former Marines. The movie does not paint any positive qualities about the U.S. Marines. It selectively amplifies the negative actions of a few, at the expense of the whole. This movie will undoubtedly have an extremely negative effect on recruiting efforts.

The overall theme paints enlisted infantry Marines as morally depraved people who quickly degrade into insanity and infighting, and completely lack honor as Desert Shield progresses into Desert Storm. Individual scenes are exceptionally offensive, to Marines and civilians alike.

While U.S. citizens have welcomed Desert Storm veterans with open arms, this movie is Hollywood's way of spitting in their face. Please boycott it, and tell others to boycott it.

*Jarhead*, starring Jake Gyllenhaal, Jamie Foxx and Peter

Sarsgaard, concerns a Marine Corps unit preparing to be deployed during the first Persian Gulf War in 1991. This movie fails on so many points, it is hard to know where to begin. The script and story are inconsistent, murky and unoriginal, with most of *Jarhead's* theme taken from earlier psychotic military epics such as *Apocalypse Now* and *Full Metal Jacket*. The movie rambles and jumps from one pointless scene to another with very little information to connect them together. The acting in *Jarhead* is average, with one vulgar performance indistinguishable from the other. The featured characters are mostly portrayed as bored, sex crazed, pornographic, maniacal Marines despondent at not being able to kill something - *anything* - after only a few months in the Gulf region. The dialogue of *Jarhead* is simplistic, and the swearing is non-stop with new lessons on how to conjugate the "F" word and its multiple variations.

Is this the same Marine Corps from World War II that actually fought an enemy for several years and died honorably at remote places such as Iwo Jima and Guadalcanal, or the continuing Hollywood portrayal of temperamental, demented Marines still suffering from the ghosts of Vietnam? In *Jarhead* we are led to believe that the soldiers are only a few steps away from an insane asylum after spending a few boring, inactive months in the war theater. *Jarhead* is a crude motion picture that has very little entertainment value whatsoever. I based my selection of which movie to see on a Friday night on the positive reviews of Roger Ebert and Richard Roeper. Where can I contact them for a refund?

Went to see this movie last night expecting a war movie about the U.S. Marine Corps. Instead I saw a movie about a bunch of nut jobs that had escaped from a mental institution. These guys were falling apart before the fighting even began. This was a movie to slam our recent efforts in the Gulf area. It was a slam on the U.S. Marine Corps, and it was a slam on the military in general. I will give you all the same advice I sent my son who is in the USMC - do not waste - *repeat*, do not waste - your money. This was a bunch of junk!

# WHO ARE THE HEROES?

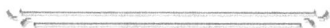

*A high school senior in Ohio named Adrienne got an English class assignment back in 2001. She had to research and write a thesis, and could pick her own topic. Adrienne dipped back into our Nation's history to a time before she was born, back to a time of national turmoil, back to the time of the war in Vietnam. Today, that long-ago conflict is a mere footnote in her history books. Who fought? Why? Who survived? Who died? Who were the heroes? From her Nation's long struggle during the war in Vietnam, Adrienne picked her topic: WHO ARE THE HEROES? An exhaustive search began. As part of her research, young Adrienne posted a notice on the website of the USMC Vietnam Helicopter Association. For the Marine Corps helicopter crews who flew and fought in Vietnam, she asked: "Who are the heroes?" The many responses included an e-mail reply from Marion Sturkey, a Marine Corps helicopter pilot in Vietnam. He wrote not of glory and valor. He never mentioned anything he did, or tried to do. Instead, he wrote of basic human virtues: commitment, loyalty, brotherly love, and a cause greater than self.*

Adrienne:

I understand you are researching a project about heroism during the war in Vietnam. I commend you for the extent of your research.

"Who are the heroes?" you ask. I had the privilege of knowing many heroes during my time in Vietnam in 1966-1967, but I doubt they are the type of men you would recognize as such. They were simply common men. Actually, "boys" would be more accurate with regard to many of them. They were not the "Follow Me!" type you may have seen in the movies. I have never heard any of them call themselves brave, although I witnessed what you would call bravery on a daily basis.

So, who are the heroes? They were the men (or "boys," many just a year or so older than yourself) who believed in each other, who relied on each other, and who sacrificed for each other. They were bound together by simple loyalty to their fellow Marines, their friends. They shared an unspoken trust and responsibility. Each knew that no matter how grave his peril, his friends would try to save him. They might fail and lose their own lives in the attempt, but we all knew that they would try. We each had the same obligation. When one of our friends was in peril, we had to try, despite the danger. We had no choice. That was the pact we made. That was our code.

Heroes were soft-spoken men like Jim McKay, a helicopter gunner. Jim had survived his scheduled time in combat and was scheduled to fly home on the night of August 8, 1966 - but that night he learned that four of his friends were cut off, surrounded, and fighting for their lives in the dark. Jim refused to leave Vietnam. He volunteered to fly on a rescue mission. His helicopter was shot down.

Heroes were men like Joe Roman, a helicopter pilot. On

January 26, 1967, he answered the plea for help from Marines trapped on a ridge in Laos. They warned him of the danger, but he disregarded the warning and flew down to attempt a rescue. He, too, got shot down. Wounded in the head and buttocks, he survived - but he never talked about it afterwards. When questioned, he would shrug and say that it was "nothing anyone else wouldn't do." He was right. Incidentally, Joe died last year. I attended his internment in Arlington National Cemetery.

There were thousands of such heroes. I am honored to have had the privilege to have served with them. Simply stated, they believed in a cause greater than themselves. They believed in each other. They knew the danger, but they also knew their responsibility and their code. They shared a brotherly love that no earthly circumstance can shatter. They, along with the fifty-eight-thousand-plus names on the Wall in Washington, D.C., are true heroes.

The heroes who survived are now in their fifties or sixties. You know them as fathers, uncles, neighbors, maybe teachers. They have jobs and families. They pay taxes and make our society function. They don't label themselves as heroes. Yet, they are American Patriots in every sense of the word, and deep down inside they still maintain that undying brotherly love for the men with whom they served in Vietnam thirty years or so ago. Without question, they are your heroes.

Warmest regards,

*Marion Sturkey*

*Adrienne got many such responses. In appreciation, she titled her thesis with the motto of the USMC Vietnam Helicopter Association: "Saepe Expertus, Semper Fidelis, Fratres Aeterni" (Often Tested, Always Faithful, Brothers Forever). In her thesis she quoted text from the book, "BONNIE-SUE: A Marine Corps Helicopter Squadron in Vietnam." She noted that, even today, "Marines religiously state 'Semper Fidelis' at the closing of letters and e-mails" sent to each other. As Adrienne now knows, the code is still alive and well.*

*Adrienne submitted her thesis. On May 1, 2001, she got the verdict. She joyfully posted another notice on the helicopter association web-site. Her notice begins: "Hey, Y'all . . . it received an 'A' with flying colors!"*

*Adrienne, who went on to attend the University of Akron, added: "This has been the most beneficial project of my high school career. I learned the most I ever could have, and will take so much with me for the rest of my life."*

**Marion Sturkey is the author of *Bonnie Sue, Warrior Culture of the U.S. Marines*, and *Murphy's Laws of Combat*. His books are highly recommended, and can be found at www.usmcpress.com.**

# NUBS
## *A Mutt, a Marine & a Miracle*

**Helena Sung**

When Marine Major Brian Dennis met a wild stray dog with shorn ears while serving in Iraq, he had no idea of the bond they would form, leading to seismic changes in both their lives. "The general theme of the story of Nubs is that if you're kind to someone, they'll never forget you - whether it be person or animal," Dennis told *Paw Nation*.

In October 2007, Dennis and his team of eleven men were in Iraq patrolling the Syrian border. One day, as his team arrived at a border fort, they encountered a pack of stray dogs - not uncommon in the barren, rocky desert that was home to wolves and wild dogs.

"We all got out of the Humvee and I started working when this dog came running up," recalls Dennis. "I said, 'Hey buddy' and bent down to pet him." Dennis noticed the dog's ears had been cut. "I said, 'You got little nubs for ears.'" The name stuck. The dog whose ears had been shorn off as a puppy by an Iraqi soldier (to make the dog "look tougher") became known as Nubs.

Dennis fed Nubs scraps from his field rations, including bits of ham and frosted strawberry Pop Tarts. "I didn't think he'd eat the Pop Tart, but he did."

At night, Nubs accompanied the men on patrols. "I'd get

179

up in the middle of the night to walk the perimeter with my weapon, and Nubs would get up and walk next to me like he was doing guard duty."

The next day Dennis said goodbye to Nubs, but he didn't forget about the dog. He began mentioning him in emails he wrote to friends and family back home. "I found a dog in the desert," he wrote in an email in October 2007. "I call him Nubs. We clicked right away. He flips on his back and makes me rub his stomach."

Every couple of weeks we'd go back to the border fort, and I'd see Nubs every time," says Dennis. "Each time, he followed us around a little more." And every time the men rumbled away in their Humvees, Nubs would run after them. "We're going forty miles an hour, and he'd be right next to the Humvee. He's a crazy fast dog. Eventually he'd wear out, fall behind, and disappear in the dust."

On one trip to the border fort in December 2007, Dennis found Nubs had been badly wounded in his left side where he'd been stabbed with a screwdriver. "The wound was infected and full of pus. We pulled out our battle kits and poured antiseptic on his wound, and force fed him some antibiotics wrapped in peanut butter." That night Nubs was in so much pain that he refused food and water, and slept standing up because he couldn't lie down. Dennis and his team left again the next day, but Dennis thought about Nubs the entire time and hoped the dog was still alive.

Two weeks later, when Dennis and his team returned, he found Nubs alive and well. "I had patched him up, and that seemed to be a turning point in how he viewed me." This time when Dennis and his team left the fort, Nubs followed. Though the dog lost sight of the Humvees, he never gave up. For two days Nubs endured freezing temperatures and packs of wild dogs and wolves, but eventually found his way to Dennis at a camp an incredible seventy miles to the south near the Jordanian border.

"There he was, all beaten and chewed up," says Dennis. "I knew immediately that Nubs had crossed through several dog territories and fought and ran, and fought and ran." Nubs jumped on Dennis, licking his face.

Most of the eighty Marines at the camp welcomed Nubs, and even built him a doghouse - but a couple of them complained, leading Dennis' superiors to order him to get rid of the dog. With his hand forced, Dennis decided the only thing to do was bring Nubs to America, so he began coordinating Nubs' rescue effort. Friends and family in the States helped, raising the five thousand dollars it would cost to transport him overseas.

Finally, it was all arranged. Nubs was handed over to volunteers in Jordan, who looked after the dog and sent him to Chicago, and then San Diego, where friends waited to pick him up. Nubs lived there with them, and began getting trained by local dog trainer Graham Bloem of the *Snug Pet Resort*. "I focused on basic obedience and socializing him with dogs, people and the environment," says Bloem.

A month later Dennis finished his deployment in Iraq and returned home to San Diego, where he immediately boarded a bus to Camp Pendleton to be reunited with Nubs. "I was worried he wouldn't remember me," said Dennis, but he needn't have worried. "Nubs went crazy. He was jumping up on me, licking my head."

Major Dennis' experience with Nubs led to a children's picture book, called *Nubs: The True Story of a Mutt, a Marine & a Miracle*, published by Little, Brown for Young Readers. They have also appeared on the *Today* show and *The Tonight Show with Conan O'Brien.*

Was it destiny that Dennis met Nubs and brought him to America? "I don't know about that," says Dennis. "It's been a strange phenomenon. It's been a blessing. I get drawings mailed to me that children have drawn of Nubs with his ears cut off. It makes me laugh."

# THE FIRST BATTLE FLAG
## *In the New War*

Sean Fontaine

Ever since Louis Daguerre accidentally took the first photograph of a person while sitting in a Paris café, it has become so that an event, no matter how significant, does not seem to be reality unless a camera makes it so. A little known event which occurred at the Pentagon on that horrific day of September 11, 2001 proved this idea to be true. A young Marine Lance Corporal raised the first battle flag on the still very unstable section of the building that was hit, and I had the honor of helping the young man get to his objective.

I was now officially out of the Corps and was a member of 20th Special Forces Group - but of course, my heart still belonged to my beloved Marine Corps. In fact, one of the chevrons I wore had the crossed rifles I had earned as a Marine. As it goes, in the midst of the chaos of the surreal scene on September 11th, a Special Forces General approached me and ordered me to accompany a young Marine who had appeared on the scene with a small U.S. flag to the roof of the Pentagon to defiantly hoist our colors for the world to see. I eagerly agreed. What an honor! It was reminiscent of the flag raising on Iwo which had become our symbol of courage and honor.

I had the gear to facilitate the posting of the colors (duct tape, 550 cord, etc), and he had the flag. We were supposed to represent the Army and Marine Corps, but above all our Nation's spirit. We climbed into a cart suspended by a cable with the operator, but the cart began to sway erratically and it became clear only two could ascend. That meant either the Marine would have to leave, or I would - so the general ordered me to tell the Marine to get out of the cart.

Every part of me recognized how wrong that was. Every part of me recognized that if *anyone* was going to raise that flag, it would be a Marine in uniform. I approached the Marine and told him what had happened. At the point of seeing his shock and disappointment, I informed him that I had served in the Corps for nine and a half years. I went on to further explain that the general had picked the wrong soldier for this mission. With a "Semper Fi," a wink and a smile, I handed him my gear and tearfully walked away from the now ascending cart.

In the darkness, the general watched as the cart climbed higher and higher. Then, at the moment it reached the top, he realized what had transpired. The Army did not get theirs that day. He angrily turned toward me, trying to make out my name, while scolding me at the same time. I apologized, and told him I was a Marine at heart – and he furiously walked away.

I saw the Lance Corporal when he returned, and he too had tears of pride in his eyes. I shook his hand and asked him if he was now a "lifer." He smiled, and answered in the affirmative.

I cannot find the words to describe how blessed I felt to be in that place on that day and time where I could help my Marine Corps one more time. So, to all my brothers and sisters both in and out of uniform... Semper Fidelis! Know that on that day, when firemen were immortalized in New York raising our flag, a young anonymous Marine did the same in Washington D.C. in the spirit of our great Nation and the United States Marine Corps.

The cameras were not there. All that remains is the image in the hearts and minds of the Marine who pridefully planted our colors and saluted while a crowd stopped, watched, saluted, fought back tears - and was newly inspired to dig through the still smoldering debris.

# KAT
# *Cold Weather, One Each*

**GySgt Edwin B. Dillard**

*I guess they were out of orange ground guides, frequency grease, and boxes of grid squares!*

It was the winter of 1991, and I was a corporal at Marine Corps Mountain Warfare Training Center in Bridgeport California. Life was good! As was the norm that time of year, we were in full swing training Marine infantry battalions in the fine art of mountaineering by conducting the Mountain Leadership Course (MLC) and Winter Survival Skills Classes. Mine was the most demanding of jobs, and I took it very seriously. In fact *everybody* depended upon me - I was a god. I was a warehouse clerk! You see, it was my duty was to issue out clothing and equipment to the Marines and students attending the local training courses.

Our warehouse staff was quite small by Marine Corps standards. There were four of us total - Sergeant West was the Warehouse Chief, and Corporal Menjarez, Lance Corporal Costa and myself were the resident "box-kickers." Together we performed our jobs admirably. Our Supply Officer, First Lieutenant Woodruff, ran a tight ship. He had been a warehouse clerk, warehouse chief, supply chief, Warrant Officer, and LDO - and with LDO promotions frozen, was the "Senior Lieutenant" in the whole Marine

Corps. He knew literally everything, and he trained us well. So well, in fact, that the four of us could issue clothing and equipment flawlessly at break neck speed. Keep in mind back in those days we did not have all the sophistication of computers, scanners, etc. All we had for documentation were the Equipment Custody Card (ECR) and the Cold Weather Card, so to speed things up we memorized the order of the equipment listed on both cards and would pre-pack the ALICE packs so that every item we pulled from it was in the same order as the entries on the cards. Like I said, we were good! It would go something like this, "Shell, glove, one pair. Inserts, glove, one pair. Cap, cold weather, one each." The whole time we were talking we were simultaneously throwing gear up on the counter as the students were reading the entries and initialing as fast as humanly possible.

It was always a pleasure issuing gear to the Officer Candidates, NROTC students, etc. because they were so disorganized compared to the regular Marines. And even more important, they were extremely unsuspecting and gullible - two dangerous ingredients when there is a "practical joker" around!

We had just finished preparing the packs for an NROTC class issue that was going later that day and Sergeant West and I went outside to see if there were any vehicles blocking the side hatch. One of the trucks that needed to be moved had an interesting passenger in the back. Underneath a thin layer of frost, in a grotesque pose, was a rather large house cat. Eyes open, tongue hanging out, it appeared to be frozen alive. After investigating the matter further we found out that a Base Maintenance man had accidentally run over his daughter's cat the night before and the frozen ground prevented him from burying it. To keep her from seeing it he did what was natural and "just threw it in the back." And you thought *Marines* were sick!

Well, we are. Without too much trouble we were able to talk him into letting us take the cat off his hands. Thankfully, he didn't ask any questions. He just raised an eyebrow and watched us carry "Stiffy" into the warehouse. We then packed the deceased feline in some snow out back and made the appropriate annotations on a few Cold Weather Cards.

At 1300 the class showed up and formed a single file line outside the side hatch. Since the issue point was small we only let four students in at a time, and the rest would stay outside the hatch and someone would come in as someone else left. Things moved quickly, as the warehouse was about sixty degrees hotter than the parking lot. We surveyed the crowd looking for a victim. The first couple of students were the class commander and a few prior service cadets - not good picks - but before long the warehouse door slammed open and in tripped a Jerry Lewis lookalike. He was tall and lanky, wore birth-control glasses, and had a bad case of "bed- head."

I tried to keep from smiling as he checked all of his pockets for appropriate identification. As he was patting himself down I ran through my spiel on what was going to happen and grabbed the special pack and issue card, and once I had his full attention I proceeded to issue his gear as fast as I could. He was trying desperately to keep up with the initials and was stuffing everything into his Waterproof Bag. I did not slow down, even when he would drop something or lose his place, and before long there was gear everywhere. He was trying to hold items under his armpit, and even his chin – and as he squared himself away I prepared for the "grand finale."

The rest of the crew was fighting back the urge to laugh as they knew what was coming next. I reached down and grabbed "Stiffy" by the tail, lifted him like a tennis racket, and dropped him on to the counter. At the same time I barked, "Kat, Cold Weather!" The cadet jumped back and looked around to see what everyone else was doing. The place was dead silent. If there were such a thing as Arctic Crickets in the world, they would have been chirping! No one else was laughing, so he grabbed Stiffy and started to initial his card. By now we were about to wet our trousers! Costa had buried his face in some poly-pro underwear, and we were all dying to laugh.

I was just getting ready to let our victim in on the joke when he spoke up, saying "Sergeant, somebody spelled Cat wrong!" With that, we all lost our bearing. Most of us were rolling around on the floor. We laughed so loudly that the Lieutenant came to investigate. He even laughed! In fact everybody was laughing... except for the cadet. As per my request he didn't tell anyone else, and I let him watch us get the next one. Before the day was over we had gotten at least six, and after our fun we gave "Stiffy" a proper burial - and talked about what we could do *next* time.

189

# THE CORPS

SgtMaj Dougherty

The Marine Corps is the only branch of the U.S. Armed Forces which recruits people specifically to fight. That is a fact. The Army emphasizes personal development (an Army of one), the Navy promises fun (let the journey begin), the Air Force offers security (it's a great way of life). Missing from all the advertisements is the hard fact that a soldier's life is to suffer and perhaps to die for his people and to take the lives of others.

Even the thematic music of the services reflects this evasion. The Army's *Caisson Song* describes a pleasant country outing over hill and dale, lacking only a picnic basket. *Anchors Aweigh*, the Navy's celebration of the joys of sailing, could have been penned by Jimmy Buffet. The Air Force song is a lyric poem of blue skies and engine thrust. All is joyful, invigorating, and safe. There are no land mines in the dales nor snipers behind the hills, no submarines or cruise missiles threaten the ocean jaunt, no bandits are lurking in the wild blue yonder.

The Marines Hymn, by contrast, is all combat with lines like, "We fight our country's battles, First to fight for right and freedom, We have fought in every clime and place where we could take a gun," and "In many a strife we have fought for life and never lost our nerve."

The choice is made clear. You may join the Army to go

to adventure training, or join the Navy to go to Bangkok, or join the Air Force to go to computer school - but you join the Marine Corps to go to war! Plus, the mere act of signing the enlistment contract confers no status in the Corps.

The Army recruit is told from his first minute in uniform that "you're in the Army now, soldier," and Navy and Air Force enlistees are sailors or airmen as soon as they get off the bus at the training center.

The new arrival at Marine Corps boot camp is called a recruit, or worse (a lot worse), but *never* a Marine. Not yet, and maybe never. He or she must *earn* the right to claim the title, and failure returns you to civilian life without hesitation or ceremony.

Recruit Platoon 2210 at San Diego, California trained from October through December of 1968, while in Viet Nam the Marines were taking two hundred casualties a week and the major rainy season and Operation Meade River had not even begun - and yet Drill Instructors had no qualms about winnowing out almost a quarter of their one hundred and twelve recruits, graduating only eighty-one.

Note that this was post-enlistment attrition. Every one of those thirty-one who were dropped had been passed by the recruiters as fit for service - but they failed the test of Boot Camp - although not necessarily for physical reasons. At least two were outstanding high school athletes for whom the calisthenics and running were child's play. The cause of their failure was not in the biceps, nor the legs - but in the spirit. They lacked the will to endure the mental and emotional strain, and so they would not be United States

Marines. Heavy commitments and high casualties notwithstanding, the Corps reserves the right to pick and choose.

History classes in boot camp? Stop a soldier on the street and ask him to name a battle of World War One. Pick a sailor at random and ask for a description of the epic fight of the Bon Homme Richard. Ask an airman who Major Thomas McGuire was and what is named after him. I am not carping, and there is no sneer in this criticism. All of the services have glorious traditions, but no one teaches the young soldier, sailor or airman what his uniform means and why he should be proud of it.

But... ask a Marine about World War One and you will hear of the wheat field at Belleau Wood and the courage of the Fourth Marine Brigade. Faced with an enemy of superior numbers who was entrenched in tangled forest undergrowth, the Marines received an order to attack which even the charitable cannot call ill-advised. It was insane. Artillery support was absent, and air support hadn't been invented yet. Even so the Brigade charged German machine guns with only bayonets, grenades, and an indomitable fighting spirit, as a bandy-legged little barrel of a Gunnery Sergeant named Daniel J. Daly, rallied his company with the shout, "Come on you sons a bitches, do you want to live forever?" He then took out three machine guns himself.

French liaison officers, hardened though they were by four years of trench bound slaughter, were shocked as the Marines charged across the open wheat field under a blazing sun and directly into the teeth of enemy fire. Their action was so anachronistic on the twentieth-century field of battle that they might as well have been swinging cutlasses - but the enemy was only human. The Boche could not stand up to the onslaught, so the Marines took Belleau Wood. The Germans, those that survived, thereafter referred to the Marines as "Tuefel Hunden" (Devil Dogs) and the French in tribute renamed the woods "Bois de la Brigade de Marine" (Woods of the Brigade of Marines).

Every Marine knows this story, and dozens more. We are taught them in boot camp as a regular part of the curriculum. Every Marine will always be taught them! You can learn to don a gas mask anytime, even on the plane en route to the war zone, but before you can wear the Eagle, Globe and Anchor and claim the title United States Marine you must first know about the Marines who made that emblem and title meaningful. So long as you can march and shoot and revere the legacy of the Corps you can take your place in line - and that line is as unified in spirit as in purpose.

A soldier wears branch service insignia on his collar, metal shoulder pins and cloth sleeve patches to identify his unit, and far too many of them look like they belong in a band. Sailors wear a rating badge that identifies what they do for the Navy. Airmen have all kinds of badges and get medals for finishing schools and showing up for work.

Marines wear only the Eagle, Globe and Anchor, together with personal ribbons and their *cherished* marksmanship badges. They know why the uniforms are the colors they are, and what each color means. There is nothing on a Marine's uniform to indicate what he or she does nor what unit the

Marine belongs to. You cannot tell by looking at a Marine whether you are seeing a truck driver, a computer programmer, a machine gunner, a cook or a baker. The Marine is amorphous, even anonymous, by conscious design.

A Marine is a Marine - period. Every Marine is a rifleman first and foremost, and a Marine first, last and always! You may serve a four-year enlistment or even a twenty plus year career without seeing action, but if the word is given you *will* charge across that wheat field! Whether a Marine has been schooled in automated supply or automotive mechanics or aviation electronics or whatever is immaterial. Those things are secondary - the Corps does them because it must. The modern battlefield requires the technical appliances, and since the enemy has them so do we - but no Marine boasts mastery of them.

Our pride is in our marksmanship, our discipline, and our membership in a fraternity of courage and sacrifice. "For the honor of the fallen, for the glory of the dead," Edgar Guest wrote of Belleau Wood. "The living line of courage kept the faith and moved ahead." They are all gone now, those Marines who made a French farmer's little wheat field into one of the most enduring of Marine Corps legends. Many of them did not survive the day, and eight long decades have claimed the rest, but their actions are immortal. The Corps remembers them, and honors what they did - and so they live forever. Dan Daly's shouted challenge takes on its true meaning. If you lie in the trenches you may survive for now, but someday you may die and no one will care. If you charge the guns you may die in the next two minutes, but you will be one of the immortals.

All Marines die, either in the red flash of battle or the white cold of the nursing home. In the vigor of youth or the infirmity of age all will eventually die, but the Marine Corps lives on. Every Marine who ever lived is living still, in the Marines who claim the title today.

It is that sense of belonging to something which will outlive our own mortality, and give people a light to live by and a flame to mark their passing.

# R.I.P. RIP

### Thomas Ripley

*I didn't have the honor of meeting John Walter Ripley until he was a Bird Colonel commanding a Regiment at Camp Lejeune - but I sure was impressed! Years later I came to find out my good friend J.C. Allen had been Ripley's AMOI at Oregon State University, and whenever he speaks of him it is with absolute reverence! This is the Colonel's eulogy, as delivered by his son and fellow Marine, Thomas Ripley.*

I would like to start my remarks with one of our father's favorite prayers - "God, Please do not let me screw this up! Amen."

As I look out in this chamber I see some of our nation's greatest warriors and patriots. Our father addressed each of you as his friend, family and fellow Marine. All are equal terms in his book. I want to thank all of you for coming to honor our father. Our family is deeply humbled by the outpouring of support and your condolences. Our father taught me that leadership is "a contact sport" in that you have to personally engage your Marines. This turnout is a testament to John Ripley's style of leadership, and for that matter his style of friendship.

My father is part of two great institutions - the U.S. Marine Corps, and the U.S. Naval Academy. I would like to take this opportunity to thank the Commandant, General

196

Conway, and the Superintendent of the Naval Academy, Vice Admiral Fowler.

When I was a newly minted second lieutenant one of my Basic School instructors told me that my father is worth a thousand men. He could see from my expression that I did not understand - a look that I *mastered* as a second lieutenant. He responded that if there was a fight and your father was going, then a thousand men would immediately join him. It appears that instructor underestimated him.

Family was always the most important thing in John Ripley's life. My siblings and I are living proof of this. We made plenty of mistakes as children, and if it were not for the bond of family none of us would have seen eighteen. Our father gave everything to us. He never had a nice car, took individual vacations, or had lavish things - all of that went to our education and betterment. He was a selfless parent. Despite all of his achievements, we always knew that we were his greatest accomplishment.

Our father loved history. A lifelong student and son of Virginia, he often closed his messages, notes and speeches with a quote from Stonewall Jackson, Jeb Stuart or Robert E. Lee. Every road trip as a child included a sudden stop to read a battlefield marker, or pause for a quick history lesson. This knowledge gave him a sense of perspective. He believed that he was located at Dong Ha for a reason, and that he had the tools and the training to blow that bridge. He passed on this perspective at every opportunity. When he would speak in public or private it was always built on stories and lessons of the past. How many of you have walked a battlefield, toured a museum, a monument or cemetery with John Ripley - the emotion was overwhelming, and was only matched by his passion for the topic. History inspired our father, and his passion was focused on the courageous acts of Marines carrying out their orders. The tactics and details of history were simply a baseline to teach us about the more important things. Our father's knowledge of the corporal, the PFC, and those overlooked acts of heroism was unmatched. It tells a great deal about John Ripley's character that he spent the time to uncover these facts, and chose to use them instead of other, more well-known examples of courage and heroism.

Faith always played a powerful role in our father's decisions. Faith is why John Ripley was always incredibly optimistic. When the odds were long John Ripley wanted to be with you, and you wanted the same thing. His parting words to my brother and I as we matriculated into VMI were "Just remember - they can't kill you." He took particular pride getting the underdog, the student athlete with the questionable academics but great leadership potential, into the Academy. John Ripley loved a good fight, and he was good at it.

There are so many incredible memories and lessons that I have from our father. I want to leave you with a few of our favorites.

Trouser pockets are not for hands. Officers with facial hair love to stand weekend duty. Always take the hard road - you will be tired, but you will be alive. There are two seats on the John Ripley train - on it, or under it. When you use the kneeler today, do not rest your bottom on the pew. Kneel or sit - not both. The term Skipper is only to be used for Commanding Officers of rifle companies in combat and ships at sea.

I want to share with you the concept of "a nickel a run." We never took normal vacations as children. In our house it was one hundred percent all the time, and this held true for vacations. We would drive all night to the beach and arrive at two in the morning. Then, in the middle of the night, we did what every normal family would do - we would go for a swim. When we went to Martha Vineyard for a day trip our father rented five bicycles, and we peddled around the island. My entire family found out first hand that it's twenty-seven miles around Martha's Vineyard. I was ten years old. While visiting Yosemite in the early spring we all had to swim across the Merced River, and when we would go skiing Dad would buy us all lift tickets at twenty dollars each. The key was to get as many runs down the hill as possible so that at the end of the day you were at "a nickel a run." I was twenty-five-years-old when I discovered that that ski resorts serve food, and have warm lodges.

In John Ripley's house everyone had a job and whining was not allowed. Idle hands are the devil's playthings. One summer our father found us watching TV in the basement. He promptly cut the power cord off the TV closest to the set so that we could not splice the wires. Many of you do not know that my father was also a dentist - after years of complaining about the fact that I had to wear braces, I demanded that they be removed. To my surprise, my father obliged and removed them that day with a pair of pliers.

You all know that our father loved being a U.S. Marine. Over and over in his notes he says the same thing - "to be a Marine is to be blessed. My emblem is the same as yours, and the same as the hundreds of thousands that went before me - they are watching me perform. I have to uphold the standards."

My father's style of leadership was summed up on a card he gave me when I was commissioned.

A Marine Leader must have:

✓ The Spirit of the Attack – Always march to the sound of the guns.
✓ Boldness - You are part of the finest fighting force in the world – act like it.
✓ A receptiveness for risk taking - Risk comes with the job, and if you are not comfortable operating with risk then you need to get into a new line of work.
✓ Endurance, mental & physical - Mental is far more important than physical.
✓ Decisiveness - Make a decision, Lieutenant!
✓ A sense of mission, a sense of duty - Mission first, Marines Always.

In recent years the Marine Corps saved our father's life - twice. The first time our 32nd Commandant, General Jones, provided a helicopter so that our father could receive a transplant. Time was short, and when John Ripley needed the Corps they came through in a way only the Corps could. The second time few people know about. The road to recovery from the transplant was a long one. Our father was struggling. There was a commotion outside his hospital room as an Army orderly tried to stop four Marines from entering. Moments later the Color Sergeant of the Marine Corps entered with the Battle Colors of the Corps. The message was simple. "Sir, the Commandant says that these colors are not to leave this room until you do." Those colors saved his life.

While stationed here in 1985 the Commandant of the Naval Academy, Leslie Palmer, died suddenly. Our father came to visit his friend prior to internment, and was shocked when he entered the room to find no honor guard with Captain Palmer. As the Senior Marine at the Naval Academy our father stood at Parade Rest by Captain Palmer for ten hours. It was only after our mother called another Marine to replace him that our father would leave his friend's side.

While serving as the Director of History for the Marine Corps our father was contacted by an officer from a regimental staff in Iraq. The regimental CP had been hit by a mortar round, and the regimental colors had blood on them and were burned in several spots. The officers question was what do we do with these Colors and how do we get new ones. Our father's response was classic John Ripley – "Nothing. They are called BATTLE COLORS."

There is little question that John Ripley was a winner, but most of you do not know the secret of our father's success. If you have ever attended one of his change of commands, a promotion or an awards ceremony, you will know that at the end he always thanked one person for his achievements - our mother, Moline. As many of you know our mother is not able to attend today's funeral service. It is said that behind every successful man is a woman that expects it. This was never more accurate than in our home. My Mom loved being married to a Marine, and she loved the Corps. I remember in Camp Lejeune when our father would come home for dinner she would politely ask him to go upstairs and change out of his uniform. I always thought this was odd. I finally figured it out. When passing out orders in the house Moline did not want to embarrass him. You see, Mom outranked Dad. Our mother, Moline, has always been the driving force behind John Ripley.

In the future I will tell my son about his grandfather, John Walter Ripley. I will tell him that John Ripley was everything that is good about being an American. He gave everything he had to his family, and he loved his wife above all things. He was driven, and eternally optimistic. John Ripley was a patriot, and he remains my hero. His one defining trait - above all others - was that John Walter Ripley is a U.S. Marine.

**The full story of Colonel Ripley's exploits can be found in the books *The Bridge at Dong Ha* by John Grider Miller and *An American Knight: The Life of Colonel John W. Ripley* by Norman Fulkerson.**

# MARINE POWs IN KOREA

*I wish this were not true - I wish ALL captured Americans had adhered to the standards of conduct expected of all our troops - but facts are facts. Just one more reason I'm glad I walked past those other recruiting offices and on down to the Marines at the end of the hall!*

In general, the performance of Americans of the other services held captive by the Communists during the Korean War shocked and disillusioned the American public. Many seemed passive and leaderless, easy marks for sophisticated "brainwashing" tactics. A distressingly large number made propaganda statements, admitting to American use of biological weapons or other absurdities. Some even opted to remain in North Korea or Red China after the armistice.

The Marine POWs fared much better. Of two hundred and twenty-one Marines captured, one hundred and ninety-four survived, a rate of eighty-seven percent (the average rate for all U.S. POWs was sixty-two percent). Marines also proved much harder to capture than troops of other services. One soldier in every one hundred and fifty who served with the Army in Korea was captured, but for the Marines it was only one in every five hundred and seventy. Many of these were captured along with the ill-fated Task Force Drysdale during the Chosin campaign, while others were from the bloody battles for the outposts in the later years of the war. Twenty Marines escaped from captivity. Others tried, and failed.

Where some POWs behaved brutally toward their fellow Americans, the Marines maintained both their sense of discipline and their adherence to the chain of command - and their smart-ass spirit. Marines in Camp Two celebrated the birthday of the Corps on November 10, 1952 by serving all hands a special cake made from stolen ingredients and topped off with a purloined bottle of rice wine. While one Marine officer succumbed to tortuous interrogation and issued a false statement (for which he was subsequently tried by USMC court-martial), the other Leathernecks kept their honor clean. As a subsequent Senate report concluded: "The United States Marine Corps did not succumb to the pressures exerted upon them by the Communists and did not cooperate or collaborate with the enemy."

Semper Fidelis - both in word, and in deed.

# *Happy...*
# ANTHONY GALE DAY!

I ask that you join me in raising a glass to Lieutenant Colonel Commandant Anthony Wayne Gale, the 4th Commandant of the Marine Corps, who was cashiered from our Corps on 18 October, 1820. Although "officially" he may not have displayed those traits which are expected of an Officer of Marines, unofficially he exhibited those qualities which are near and dear to many of our hearts.

You should also know that our 4th CMC is the only one whose picture is *not* posted in the hallowed halls of HQMC... imagine that! Methinks I served with a few Marine Corps officers who more than matched Gale's escapades. It might have even been me...

Lieutenant Colonel Gale was born in Dublin, Ireland on 17 September 1782. Fewer records survive concerning him than any other Commandant, but it is known that when he was commissioned a second lieutenant on 2 September 1798 he was one of the first officers commissioned after the reestablishment of the Marine Corps.

Thereafter he fought, in fairly quick succession, the French, the Barbary pirates, the British, and one of his Navy shipmates. The last encounter, involving an affront to the Corps, brought about the naval officer's sudden demise and met with Commandant of the Marine Corps William Ward

Burrow's approval for Gale's defense of his Corps' honor. As the story goes, Gale was Ship's Company Commander aboard *USS Ganges* in November of 1799 when Navy Lieutenant Allen McKenzie had one of the Marines put in irons without first consulting Gale. When Gale inquired about the incident, McKenzie called him a "rascal."

The rest of the story is related in correspondence by Commandant Lieutenant Colonel Burrows:

*"The Captain took no notice of the business and Gale got no satisfaction on the cruise. The moment he arrived he called (McKenzie) out and shot him. Afterwards, politeness was restored."*

McKenzie died of his wounds, and Burrows went on to say, "It is hoped that this may be a lesson to the Navy Officers to treat the Marines, as well as their Officers, with more respect."

Unfortunately for Captain Gale, increasing rank brought other difficulties not resolved so directly. In 1815 Burrows' successor as Commandant, Lieutenant Colonel Franklin Wharton, was charged by Congress with over-spending on the construction of Marine Barracks Philadelphia. He in turn accused the Commandant of the Barracks, Anthony Gale, of building "extravagant" officers' quarters. Gale was ordered to stand before a Court of Inquiry, but was exonerated. It was shortly after this that Wharton was again called to account to Congress. This time he was accused of fleeing Washington, rather than leading his Marines in the Battle of Bladensburg.

When convened, his Court Martial consisted of three Navy Captains and one Captain of Marines - Captain Anthony Gale. The Court decided that Marines ashore were subject to Army, rather than Navy, Courts under the Articles of War, and the charges were ultimately dropped. His duties on his Commandant's Court Martial complete, Gale was promoted to Major and transferred to command Marine Barracks New Orleans. Soon afterwards a letter to the Secretary of the Navy reported that Navy officers had "frequently seen Major Gale intoxicated at New Orleans and that his associates were of such a description and his habits of such a nature as to prevent the respectable officers of that station from having any social or friendly intercourse with him."

Daniel T. Patterson, Commander of the New Orleans Naval Station, wrote to the Secretary, "It is reluctantly and with extreme regret that I have again to address you relative to the Marines of this station, but longer to remain silent would be to neglect my duty. The Non-Commissioned Officers and Privates are, without exception, the most depraved, abandoned, and drunken set of men ever collected together." While Gale was preparing to go to Washington to answer the preceding charges, Commandant Wharton died. At his Court Martial Gale was found not guilty and returned to duty, and as he was the next senior officer in the Marine Corps he was nominated to become Commandant.

Despite the vigorous protests and political maneuvering of the Paymaster of the Marine Corps and Major Archibald Henderson (each of whom felt themselves better qualified for the position), on 5 March 1818 Gale was confirmed as the fourth Commandant of the Marine Corps. With it came promotion to Lieutenant Colonel.

His tenure was to be brief. Soon came troubles with Navy Secretary Thompson, who frequently countermanded Gale's orders in a humiliating manner. Finally, Lieutenant Colonel Gale courageously submitted a letter analyzing the proper division of function between himself and the Secretary, and respectfully pointed out the impossibility of his position. This official reaction to infringements of his authority he paralleled by unofficial retreats to alcohol. On 18 September 1820 he was arrested and charged with offenses of an alcoholic and related nature. The first charge was that he was publicly intoxicated in the city of Washington on six specified dates - during the month of August. There were also several specifications under the charge of "Conduct Unbecoming an Officer." First, that he had visited a house of prostitution near the barracks, "in an open and disgraceful manner" and second, that on 1 September he had, before witnesses, called the Paymaster of the Marine Corps "a damned rascal, a liar, and a coward." Other charges concerned his breaking house arrest and maintaining a Marine as a personal servant.

Gale's unsuccessful defense was temporary insanity. He was cashiered from the Marine Corps on 18 October 1820, leaving forty-six other officers on active duty in the Corps, and Archibald Henderson succeeded him as Commandant. From Washington Gale went first to Philadelphia, where he spent several months in hospitals, and then took up residence in Kentucky. Armed with proof that he had been under the strain of temporary mental derangement while Commandant, he spent fifteen years attempting to have his court-martial decision reversed. Eventually, in 1835, the government partially cleared him and awarded a stipend of fifteen dollars a month, which was later increased to twenty-five and continued until his death in 1843 in Stanford, Lincoln County, Kentucky.

Today no trace of his grave exists, and Anthony Wayne Gale is the only Commandant of whom the Marine Corps has no portrait.

**The preceding account was partially compiled from *The U.S. Marine Corps Story* by J. Robert Muskin, and *U.S. Marines: 1775-1975* by Brigadier General Edwin Simmons USMC (Retired).**

# OLD DOGS

*Here's an important "Life Rule" every punk should know: Don't pick a fight with an old Marine, because if he's too old to fight, he'll just kill you. Here's the proof:*

A seventy-one-year-old retired Marine named John Lovell opened fire on two robbers at a Plantation, Florida sub shop late on a Wednesday evening, killing one and critically wounding the other. Lovell, a former pilot for two presidents, doesn't drink, doesn't smoke and works out every day.

According to Plantation police, two masked gunmen came into the Subway just after eleven PM. There was a lone diner - Lovell - who was finishing his meal. After robbing the cashier, the two men attempted to shove Lovell into a bathroom and rob him as well. They got his money, but then Lovell pulled his handgun and opened fire, shooting one of the thieves in the head and chest and the other in the head. When police arrived they found one of the men in the shop, and K-9 units found the other in the bushes of a nearby business. They also found cash strewn around the front of the sandwich shop. Both men were taken to Broward General Medical Center where one died, and the other was in critical but stable condition.

A longtime friend of Lovell was not surprised to hear what happened. "He'd give you the shirt off his back, and he'd be mad if someone tried to take the shirt off your back,'" he said. Lovell worked as a pilot for the Marines, flying former Presidents John F. Kennedy and Lyndon B. Johnson, and later worked as a pilot for Pan Am and Delta.

He was not charged, authorities said, because "he was in fear for his life. These criminals ought to realize that most men in their seventies have military backgrounds and aren't intimidated by idiots.

The only thing the perpetrators could be charged with is participating in an unfair fight - one seventy-one-year young Marine against two punks. Two head shots and one center-of-mass shot – that's good shooting! That'll teach them not to get between a Marine and his meal. Don't you just love a story with a happy ending?

**Note: Florida law allows properly licensed, law-abiding citizens to carry a concealed weapon.**

# ONE HILL, ONE MARINE

On November 15, 2003 an eighty-five-year-old retired Marine Corps colonel died of congestive heart failure at his home in La Quinta, California, southeast of Palm Springs. He was a combat veteran of World War II - reason enough to honor him - but this Marine was a little different. This Marine was Mitchell Paige.

It's hard today to envision - or, for the dwindling few, to remember - what the world looked like on October 26, 1942. The U.S. Navy was not the most powerful fighting force in the Pacific, not by a long shot, so the Navy basically dumped a few thousand lonely American Marines on the beach on Guadalcanal and high-tailed it out of there.

You Navy guys can hold those letters. Of course Nimitz, Fletcher and Halsey had to ration what few ships they had. I've written separately about the way Bull Halsey rolled the dice on the night of November 13, 1942, violating the stern War College edict against committing capital ships in restricted waters and instead dispatching into the Slot his last two remaining fast battleships, *South Dakota* and *Washington*, escorted by the only four destroyers with enough fuel in their bunkers to get them there and back.

Those American destroyer captains need not have worried about carrying enough fuel to get home. By eleven PM, outnumbered better than three-to-one by a massive Japanese task force driving down from the northwest, every one of those four American destroyers had been shot up, sunk or set aflame. And while *South Dakota* - known throughout the fleet as a jinx ship - had damaged some lesser Japanese vessels, she continued to be plagued with electrical and fire control problems.

"*Washington* was now the only intact ship left in the force," writes naval historian David Lippman. "In fact, at that moment *Washington* was the entire U.S. Pacific Fleet. She was the only barrier between (Admiral) Kondo's ships and Guadalcanal. If this one ship did not stop fourteen Japanese ships right then and there, America might lose the war."

On *Washington's* bridge, Lieutenant Ray Hunter had the conn. He had just seen the destroyers *Walke* and *Preston* "blown sky high." Dead ahead lay their burning wreckage. Hundreds of men were swimming in the water, and the Japanese ships were racing in.

"Hunter had to do something. The course he took now could decide the war," Lippman writes. "'Come left,' he said, and *Washington's* rudder change put the burning destroyers between her and the enemy, preventing her from being silhouetted by their fires.

The move made the Japanese momentarily cease fire. Lacking radar, they could not spot the American battleship behind the fires. *Washington* raced through burning seas, as dozens of destroyer men were in the water clinging to floating wreckage. "Get after them, *Washington!*" one shouted.

Sacrificing their ships by maneuvering into the path of

torpedoes intended for *Washington*, the captains of the American destroyers had given China Lee one final chance.

Blinded by the smoke and flames, the Japanese battleship *Kirishima* turned on her searchlights, illuminating helpless *South Dakota*, and opened fire. Finally, as her own muzzle blasts illuminated her in the darkness, Admiral Lee and Captain Glenn Davis could positively identify an enemy target.

*Washington's* main batteries opened fire at precisely midnight. Her radar fire control system functioned perfectly, and during the first seven minutes of November 14, 1942 the "last ship in the U.S. Pacific Fleet" fired seventy-five of her sixteen-inch shells at the battleship *Kirishima*. Aboard *Kirishima*, it rained steel. At 3:25 AM, her burning hulk officially became the first enemy sunk by an American battleship since the Spanish-American War. Stunned, the Japanese withdrew. Within days, Japanese commander Isoroku Yamamoto recommended the unthinkable to the emperor - withdrawal from Guadalcanal.

But that was still weeks in the future. We were still with Mitchell Paige back on the Godforsaken malarial jungle island of Guadalcanal, placed like a speed bump at the end of the long blue-water slot between New Guinea and the Bismarck Archipelago - the very route the Japanese Navy would have to take to reach Australia.

On Guadalcanal the Marines struggled to complete an airfield. Yamamoto knew what that meant, and no effort would be spared to dislodge these upstart Yanks from a position which could endanger his ships. Before long, relentless Japanese counterattacks had driven supporting U.S. Navy ships from inshore waters. The Marines were on their own.

As Platoon Sergeant Mitchell Paige and his thirty-three riflemen set about carefully emplacing their four water-cooled .30-caliber Brownings, manning their section of the thin khaki line which was expected to defend Henderson Field against the assault of the night of October 25, 1942, it's unlikely anyone thought they were about to provide the definitive answer to that most desperate of questions - how many able-bodied U.S. Marines does it take to hold a hill against two thousand desperate and motivated attackers?

Nor did the commanders of the mighty Japanese Army, who had swept all before them for decades, expect their advance to be halted on some God-forsaken jungle ridge manned by one thin line of Yanks in khaki - but by the time the night was over, the 29th (Japanese) Infantry Regiment had lost 553 killed or missing and 479 wounded among its 2,554 men. The 16th (Japanese) Regiment's losses are uncounted, but the 164th's burial parties handled 975 Japanese bodies, so the American estimate of 2,200 Japanese dead is probably too low.

You've already figured out where the Japanese focused their attack, haven't you? Among the ninety American dead and seriously wounded that night were the men in Mitchell Paige's platoon. Every one of them. As the night of endless attacks wore on Paige moved up and down his line, pulling his dead and wounded comrades back into their foxholes and firing a few bursts from each of the four Brownings in turn, convincing the Japanese forces down the hill that the positions were still manned. The citation for Paige's Medal of Honor picks up the tale:

*"When the enemy broke through the line directly in front of his position, P/Sgt. Paige, commanding a machinegun section with fearless determination, continued to direct the fire of his gunners until all his men were either killed or wounded. Alone, against the deadly hail of Japanese shells, he fought with his gun and when it was destroyed, took over another, moving from gun to gun, never ceasing his withering fire."*

In the end, Sergeant Paige picked up the last of the forty-pound, belt-fed Brownings - the same design which John Moses Browning famously fired for a continuous twenty-five minutes until it ran out of ammunition, glowing cherry red, at its first U.S. Army trial - and did something for which the weapon was never designed. Paige walked down the hill toward the place where he could hear the last Japanese survivors rallying to move around his flank, with the belt-fed gun cradled under his arm, firing as he went - and the weapon did not fail.

Coming up at dawn, battalion executive officer Major Odell M. Conolley was first to discover the answer to our question of "how many able-bodied Marines does it take to hold a hill against two regiments of motivated, combat-hardened infantrymen who have never known defeat?

On a hill where the bodies were piled like cordwood,

Mitchell Paige alone sat upright behind his .30-caliber Browning, waiting to see what the dawn would bring. One hill - one Marine.

But in the early morning light, the enemy could be seen a few yards off, and vapor from the barrels of their machine guns was clearly visible. It was decided to try to rush the position. For the task, Major Conolley gathered together three enlisted communication personnel, several riflemen, a few company runners who were at the point, together with a cook and a few messmen who had brought food to the position the evening before. Joined by Paige, this ad hoc force of seventeen Marines counterattacked at 5:40 AM, and discovered that the extremely short range allowed the optimum use of grenades. They cleared the ridge.

And that's where the unstoppable wave of Japanese conquest finally crested, broke, and began to recede. On an unnamed jungle ridge on an insignificant island no one had ever heard of called Guadalcanal.

But who remembers today how close-run a thing it was - the ridge held by a single Marine, in the autumn of 1942? When the Hasbro Toy Company called some years back, asking permission to put the retired colonel's face on some kid's doll, Mitchell Paige thought they must be joking - but they weren't. That's his mug, on the little Marine they call "G.I. Joe."

And now you know.

# ONCE A MARINE

**James Taylor**

As I came out of the supermarket that sunny day, pushing my cart of groceries towards my car, I saw an old man with the hood of his car up and a lady sitting inside the car with the door open. The old man was looking at the engine. I put my groceries away in my car and continued to watch the old gentleman from about twenty-five feet away. I then saw a young man in his early twenties with a grocery bag in his arm, walking towards the old man. The old gentleman saw him coming too and took a few steps towards him. I saw the old gentleman point to his open hood and say something. The young man put his grocery bag into what looked like a brand new Cadillac Escalade, and then turn back to the old man and I heard him yell at the old gentleman saying, "You shouldn't even be allowed to drive a car at your age." And then with a wave of his hand, he got in his car and peeled rubber out of the parking lot.

I saw the old gentleman pull out his handkerchief and mop his brow as he went back to his car and again looked at the engine. He then went to his wife and spoke with her, and appeared to tell her it would be okay. I had seen enough, and approached the old man. He saw me coming and stood straight and as I got near him I said, "Looks like you're having a problem." He smiled sheepishly, and quietly nodded his head. I looked under the hood myself and knew that whatever the problem was, it was beyond me. Looking around I saw a gas station up the road and told the old gentleman that I would be right back. I drove to the station and went inside and saw three attendants working on cars. I approached one of them and related the problem the old man had with his car and offered to pay them if they could follow me back down and help him.

The old man had pushed the heavy car under the shade of a tree and appeared to be comforting his wife. When he saw us he straightened up and thanked me for my help, and as the mechanics diagnosed the problem (overheated engine) I spoke with the old gentleman. When I shook hands with him earlier he had noticed my Marine Corps ring and had commented about it, telling me that he had been a Marine too. I nodded and asked the usual question, "What outfit did you serve with?" He had mentioned that he served with the first Marine Division at Tarawa, Saipan, Iwo Jima and Guadalcanal. He had hit all the big ones, and retired from the Corps after the war was over.

As we talked we heard the car engine come on and saw the mechanics lower the hood. They came over to us as the old man reached for his wallet, but was stopped by me and I told him I would just put the bill on my AAA card. He still reached for the wallet and handed me a card which I assumed had his name and address on it, and I stuck it in my pocket. We all shook hands all around again and I said my goodbyes to his wife. I then told the two mechanics that I would follow them back up to the station.

Once at the station I told them that they had interrupted

their own jobs to come along with me and help the old man, and said I wanted to pay for the help - but they refused to charge me. One of them pulled out a card from his pocket looking exactly like the card the old man had given to me. Both of the men then told me that they were Marine Corps Reserves. Once again we shook hands all around, and as I was leaving one of them told me I should look at the card the old man had given to me and I said I would and drove off.

For some reason I had gone about two blocks when I pulled over and took the card out of my pocket and looked at it for a long, long, time. The name of the old gentleman was on the card in golden leaf, and under his name it said, "Medal of Honor Society." I sat there motionless, looking at the card and reading it over and over. I looked up from the card and smiled to no one but myself, and marveled that on this day four Marines had all come together because one of us needed help. He was an old man alright, but it felt good to have stood next to greatness and courage, and it was an honor to have been in his presence.

# TEAM BOX SCORE

### Dick Camp

*Keep this mission in mind when you read the NEXT story, and think about what these Marines went through defending the "free speech rights" of charlatans and "posers."*

Marine Captain David F. Underwood's Sikorsky UH-34D Seahorse helicopter of Marine Medium Helicopter Squadron 163 was refueling when he heard the direct air support center (DASC) announce over the radio that a reconnaissance team was surrounded by North Vietnamese Army (NVA) regulars and needed to be extracted immediately. One CH-46 Sea Knight helicopter already had been shot out of the zone while trying to reach the embattled Marines, and the DASC reported that the 46 had taken heavy fire.

Captain Underwood contacted the DASC. "I'll give it a try if you want me to do it," he radioed.

"Dave Underwood felt like we could accomplish the mission," his wingman, Captain Carl E. Bergman reported. "He knew the gravity of the situation and the seriousness of getting it done rather than waiting for another 46. The team just couldn't wait that long, so Dave decided to go in."

As helicopter gunships and fixed-wing aircraft bombed and strafed the area Underwood and his copilot, Captain Tom Burns, flew their "34" through a hail of North Vietnamese automatic-weapons fire. They set the helicopter down on a little peak, fully exposed to enemy soldiers who were blasting them at point-blank range.

"We were taking just unbelievable fire at this point," Underwood exclaimed. "All the glass was blown out of my instrument panel. The windshield was blown out. You could hear the bullets going through the cockpit like bees!"

Underwood and his crew were in mortal danger, taking heavy fire while the heavily burdened reconnaissance team struggled to reach them.

Second Lieutenant Terrence C. Graves was one of the first of his eight-man 3rd Force Reconnaissance Team 2-1, call sign "Box Score," to jump down from the bed of the stopped deuce-and-a-half truck. He quickly led his men into the brush where they formed a small perimeter and waited, alert for sounds that indicated their covert insertion had been discovered.

The truck gathered speed and continued on its way, trying to give the impression that nothing unusual had occurred. The team believed that the enemy was watching the roads closely, so after several minutes Graves gave a signal and they silently moved out, well aware this "Indian Country" was alive with NVA.

The team had been thoroughly briefed before leaving the Force Recon area at Dong Ha, and was to "conduct reconnaissance and surveillance in their assigned zone to determine enemy activity." In accordance with standard procedure they were to use supporting arms to engage the enemy, and were also to attempt to capture a prisoner. Secondarily, the team was to plot helicopter landing zones (LZs) for future operations. They were cautioned to pay particular attention to trails to determine if the enemy used them.

Team Box Score "broke brush" for the first day, slowly moving farther into their patrol area approximately six miles northwest of Dong Ha in Quang Tri Province just south of the Demilitarized Zone (DMZ). They saw "lots of enemy activity, primarily footprints on the trails." The seven enlisted members had worked together on several missions and were comfortable in the enemy's backyard, but although this was his fourth patrol it was Graves' first experience as a patrol leader since the company believed in first putting perspective patrol leaders in subordinate positions to gain experience.

The team's two corporals, Robert B. Thomson and Danny M. Slocum, were "extremely well-qualified reconnaissance Marines."The other five men, Lance Corporal Steven E. Emrick, Hospital Corpsman Third Class Stephen B. Thompson and Privates First Class James Earl Honeycutt, Adrian S. Lopez and Michael P. Nation, were typical - all volunteers, physically fit, well-trained and highly motivated Marines.

By late afternoon the team had reached its reconnaissance zone and scouted for a nighttime harbor site. "After dark," Nation recalled, "we found a brush-covered area that offered good concealment." They established a circle, and immediately planned close-in artillery targets in case of attack. "Every two hours we would rotate the watch, which allowed everyone to get some much needed rest. The night passed slowly, as it does when you're on patrol, but nothing out of the ordinary was observed."

By dawn the team had moved west through thick scrub toward hills covered with waist-high elephant grass. "After crawling through some very low brush," Nation said, "we could see a well-used trail just across a stream in front of us." The team paused. "Suddenly we heard Vietnamese voices quite a ways away, so we all got down, moved to the side of the brush line and waited to see if they came closer."

When they didn't, Graves decided to move the team to a better position to observe the enemy and possibly capture one. "We crossed the little streambed and crawled up the hill to a bomb crater where we formed a 360," Nation explained. "That's when I spotted five NVA carrying packs and rifles coming down the path toward us." The Marine patrol was caught in the middle of two NVA units.

They didn't have much of a choice but to lay an ambush, so Graves passed the word to execute their ambush drill. "We peeled off and set up a hasty ambush alongside the trail as best we could because the brush was only two, maybe three, feet high," Nation described. "When it came my turn there was no cover, so Honeycutt and I jumped into a ten-foot deep, steep-sided bomb crater. All I could see was sky."

The NVA, now numbering seven, continued down the hillside trail. "Four of us (Graves, Lopez, Thomson and Slocum) moved up the hill to ambush them," Slocum recalled. The NVA approached the kill zone. "One guy got about fifteen to twenty feet away from me and kind of looked over my way, and there was another one that just seemed to pop right out of the ground."

Nation could not see the second enemy soldier. "I think he may have seen me, so I had to open fire. I shot him straight through the head with my M14, and then everybody opened fire... Thomson with an M79 grenade launcher, the others with small arms and a couple of hand grenades."

The team stopped firing and prepared to check out the kill zone, and as Slocum stood up to move to another position he was wounded. "All of a sudden one or two rounds were fired," Nation recalled. "At first, I thought he was hit in the head because he fell back, but then he sat up." The team returned fire, and then Graves and Thomson crept into the kill zone to check out the bodies.

"They came back a few minutes later with a pack, diary and a few odds and ends," "Doc" Thompson said. "All the NVA were dead. The trail was just one big mass of blood." Thompson treated Slocum for "a minor wound in the upper right thigh... a 'through and through' wound that basically took out some skin. I put a couple of battle dressings on it and offered Danny morphine, but I didn't recommend doing that because it would slow some of his senses. He said, 'That's fine, Doc. If I need it, I'll let you know.' I believe that decision saved his life." Although relatively minor, the wound was serious enough to prevent Slocum from continuing on the patrol. Graves requested a medevac helicopter and then ordered the team to move to the top of the hill.

"As we started moving up, we got pinned down by automatic-rifle fire," Nation recalled, "kinda like the movies with the rounds bouncing off the ground." The team returned fire, allowing the Marines to move up the hill where they formed a circle and waited for the medevac helicopter.

"The NVA seemed to be getting closer, pretty much from all directions," Nation explained. "I could see several. Honeycutt and I started shooting... I think he got three, and I got one... but we started getting rounds in, enough to make us want to stay down low."

Graves and Emrick worked the radio, directing artillery and air support. "The fire was so heavy," HM3 Thompson said, "Lieutenant Graves would sit up and see where the round hit and lay back down and call for adjustment." He also directed Huey gunships that had responded to the call, "Troops in contact!"

Graves passed the word that the medevac "bird" was coming in, and Nation laid out an air panel to mark their location. "We started toward the hovering helicopter," Thompson recalled, "when all hell broke loose!"

The NVA focused their fire on the bird. "It was being riddled with machine-gun fire," Thompson said. "It looked to me like the copilot and the gunner were hit." Before the team could reach it, the damaged helicopter lifted out of the zone.

"The area was obscured by smoke from rockets that the Hueys and F-8s (jets) were firing," Captain Underwood remembered. "I saw the 46 enter the area and momentarily reappear through the smoke coming back out."

As the damaged helicopter took off, automatic-weapons fire raked the team's position. "The lieutenant, Emrick and Thomson all got hit," Nation recalled. "I remember the lieutenant was the first to yell that he got hit." Honeycutt bandaged the minor wounds Graves received in the upper thigh. The other two were wounded seriously.

"Thomson was hit in the lower waist," Doc Thompson remembered. "He said, 'I'm blacking out, Doc. I'm blacking out.' Then he passed out on me, and I think at that moment he died, although I started closed chest cardiac massage and mouth-to mouth resuscitation."

Nation tried to help Emrick. "When I flipped him over, he said, 'Get the radio off...' and that's the last thing he said." Nation administered mouth-to-mouth resuscitation because Lopez still could feel a pulse. After being bandaged, Graves limped back to work. "He directed air strikes," Thompson said, "and kept up a small base of fire to give us some protection."

The radio nets at the reconnaissance company's command post were filled with the team's urgent requests for assistance. Plaintive calls galvanized the entire spectrum of support... fixed- and rotary-wing aircraft, artillery tubes and an infantry reaction force. Lieutenant Colonel William D. Kent, the commanding officer, closely monitored events but "felt absolutely helpless." Everything he could do for the team was being done.

Captain Underwood was some distance away when he heard about the unsuccessful extraction attempt. "The 46 took heavy fire and couldn't bring the team out," he explained, "so I left my wingman (in orbit) and went down on the deck to meet the Huey gunship to lead me into the zone." Underwood needed a guide because the zone was obscured almost totally by smoke from six gunships and two F8U fighter/bombers that were mercilessly pounding the NVA positions. Captain Bobby F. "Gabby" Galbreath, a friend of Underwood's flying a UH-1 from Marine Observation Squadron 6, volunteered to lead him in. "Just follow me, and when I break, the zone will be right underneath me," Galbreath radioed.

"I followed him in, going flat out," Underwood said. "As he broke left, I button hooked and brought my aircraft into a low hover on top of the ridge." As Underwood's helicopter started its descent, it came under intense automatic-weapons fire. "I could actually see the NVA blasting away with AK47s... unbelievable fire... anything except a 34 would have been blown out of the sky. My rotor wash was pushing the elephant grass down, and I tried to spot where the guys were because I couldn't see them. I air-taxied down the ridge until we finally spotted one of them half-hidden in the grass, dragging a guy who'd been wounded."

The team struggled with the casualties. Doc Thompson and PFC Honeycutt dragged Corporal Thomson, while PFCs Nation and Lopez handled Lance Corporal Emrick as Lieutenant Graves and Corporal Slocum provided covering fire. "We couldn't stand up because the fire was still coming in on us and the grass was so short," Nation recalled. "You had to just kind of kneel down and pull them, while trying to keep them breathing."

Underwood remained in the zone more than three minutes under heavy fire. "I told my crew chief to lay down suppressive fire and to try and hurry the team up,"

Underwood said, "but we were going to stay as long as we could to get everyone on board."

Corporal Al Mortimer, the crew chief, frantically gestured for the team to hustle. "Two recon members brought up one man (Thomson). They loaded him on board, and then the corpsman came in. He started pounding on the man's heart, trying to keep it going."

Nation, Lopez and Honeycutt struggled to load Emrick "because he was so heavy," according to Nation. After sliding the mortally wounded man into the cabin, Honeycutt jumped off the 34's step to assist Graves and Slocum in trying to suppress the NVA fire. Lopez also jumped to the ground, but a fast-thinking Mortimer stopped him. "I grabbed him by the collar and was helping him in when he got shot in the leg. 'I'm hit!' he yelled. At that time we took off and were in the air by the time I pulled him in the plane," Mortimer recalled.

Thomson and Emrick lay on the helicopter's bloody deck. The corpsman, hunched over Thomson, massaged his heart while Nation worked on Emrick. After a few minutes, Nation realized that "Emrick was gone, and I gave up on him." He then started working on Lopez who was unconscious from loss of blood.

"The doc gave me his Ka-Bar, and I cut his pants leg open and pressed a bandage on the wound to get the blood to stop. It was just gushing all over the bottom of the chopper." The bullet had severed Lopez's ephemeral artery, ricocheted into his abdominal cavity and exited through his right hip.

Bullets suddenly riddled the helicopter. "The whole side

of the chopper seemed to be coming in on us," Nation said. "Some of the stuff hit me in the face." Honeycutt was just climbing in, Slocum was to one side, and Graves was a few feet away. "The lieutenant was screaming at the top of his lungs," Nation exclaimed. "'Get out! Get out!' And he just waved at the chopper pilot to get the hell out of there 'cause he can see that the fuel tanks had been ruptured." Thompson was convinced that "Terry (Graves) probably knew that he was not going to live at that point. He knew that the chopper was hit so badly that the extra weight would have kept it from taking off."

Thompson could see that "the pilot was working as hard as he could to get the chopper in the air. It was severely hit."As the helicopter lifted out of the zone, Underwood saw Honeycutt jump out. "I called my crew chief and asked him how many men we had aboard. I was informed that we only had five... that two had jumped out, obviously in search of a third one who was wounded. At this point there was nothing I could do."

Underwood stayed at treetop level and "poured the coal" to his aircraft. "The closest place was Delta Med ("Delta" Company, 3dMedical Battalion) at Dong Ha," Underwood said. "I landed there and shut the helicopter down."

Nation recalled, "The crew shouted for us to get out. There was fuel running out all over the place." The helicopter had taken twenty hits, the majority in the cockpit. Underwood said, "I could not have flown it anymore. In fact, it had to be lifted out."

The wounded were rushed into surgery. "They worked as

hard as they could on Thomson," a corpsman said, "but couldn't keep him alive." The doctors stabilized Lopez and evacuated him to a larger hospital at Da Nang, but he died the next day.

Nation sat outside, "not knowing what the hell to do." Within five minutes someone from his unit arrived and started pumping him for information. "'Tell us what happened,' and I'm telling them the best I can so we can get a reaction team together to go back and get the others."

As the rescue attempt was planned, the other three members left behind were fighting for their lives. Graves, Honeycutt and Slocum scrambled away from the LZ to the top of the hill, "so we wouldn't catch so much incoming and waited for the next helicopter," Slocum said. It was not long in coming.

Captain Bergman followed a gunship toward the new LZ. "We went in low," his copilot, Captain Ed Egan, recalled. "The ground was pretty well obscured by the smoke from WP (white phosphorus), so it was just about IFR (instrument flying) down on the deck." They couldn't find the zone and were forced to make three more attempts before finding it.

"As soon as we sat down, I could see them at my eleven o'clock about fifteen meters away," Egan said, "but they didn't make any move to get in the aircraft. Just about that time, we came under fire. It was so close and so loud that I thought our gunner had opened up."

The gunfire was so loud that Bergman did not hear his copilot say that he saw the team, and he started to lift out of the zone. At point, "the aircraft gunner got on the ICS (intercom)," Egan related, "and said that the crew chief had been shot in the shoulder and was bleeding badly. I looked down between the seats, and I saw blood all over the cabin deck and the crew chief lying there."

With a crewman down, the aircraft badly damaged and leaking fuel and hydraulic fluid, one of the radios shot up and both windows on the copilot's side shot out, Bergman could do no more. He headed for Dong Ha. "I had a decision to make, and I made it," Bergman said regretfully. "I didn't accomplish the mission I was trying to do."

The three men on the ground "moved to the south side of the hill so we wouldn't catch as much incoming," Slocum recalled. "We got all the packs and stuff together that we were going to take. Lieutenant Graves got a radio, I got one, and Honeycutt got most of the weapons."

With the NVA closing in on the three men, Captain Galbreath decided to attempt a rescue with his Huey gunship. "Don't go in there," Underwood warned. "The fire is too intense. You'll never make it."

"Naah," Galbreath responded nonchalantly, "I'm going to go in and try to get them."

Galbreath piloted the vulnerable aircraft through intense machine-gun and small-arms fire. "The NVA were really putting some rounds in it," Slocum said. "It never quite touched the hill, just kind of hovered about a foot off the ground." The three recon Marines ran for their lives and scrambled aboard. "We all got on and it took off. I'd say not even five meters off the ground." Enemy fire riddled the Huey as Galbreath tried to gain airspeed. "I saw the copilot slump over as rounds came through the rear section of the chopper, cutting up people," Slocum recalled. "I also think the lieutenant (Graves) got hit again."

The damaged Huey lurched out of control. "It was completely spastic," Slocum said, "and crashed on its side across the river, about fifty meters from the bank, right above a bomb crater." The tremendous impact hurled everyone together in a tangled heap. "I was on top of the pile, so I was able to shimmy out." He jumped to the ground and found "one of the pilots stretched out on the ground, semiconscious."

Slocum could see a line of fifteen to twenty enemy soldiers closing in. "One of them was yelling orders, not paying attention to anything. So, I asked the pilot if he had a pistol. He said, 'No, but he had a carbine in the cockpit.' I couldn't find it, so I climbed up on the chopper and tried to get the machine gun. Just as I got my hands on it the gooks opened up, so I jumped off and landed about five to six meters from the helicopter. I froze near the wreckage of a rocket pod, hoping they wouldn't see me." He heard the NVA soldiers firing one or two rounds at a time.

Slocum headed downstream and spent the night dodging friendly artillery harassing and interdiction (H&I) fire. "The next morning I got up at first light and followed a trail. As I got to the top, I noticed a gook about twenty meters away." Slocum quickly backtracked until the trail crossed a stream. "I went up the stream for 100 to 150 meters and crawled up the bank." The NVA were close enough that "I could hear them talking. They seemed to be coming toward me."

Slocum inched his way through the brush to the top of the hill where he could see helicopters and hear the sounds of a firefight.

Slocum was unaware that late the previous afternoon 2d Platoon, Company B, 1st Battalion, Fourth Marine Regiment was helo-lifted into the area to aid the recon team. Known as a "Sparrowhawk," the small reaction force rushed to the scene of the crash. As the platoon approached the downed helicopter, the NVA suddenly opened fire from three sides with small-arms and automatic-weapons fire. Corporal William A. Lee, the platoon radio operator, was struck in the head and chest and fell mortally wounded.

The NVA closed in and threw Chicom (Chinese communist) grenades into the ranks of the advancing Marines, slightly wounding four. Under threat of being overwhelmed, the platoon leader wisely withdrew his force 150 meters to the southeast and called in mortar and artillery defensive fires throughout the night. An AC-47 gunship, nicknamed "Puff the Magic Dragon," circled overhead providing illumination and fire support from its six-barreled, rotating 7.62 mm miniguns, which could cover a football field with one round in one minute.

Early the next morning, the rest of the company joined the platoon. The combined force reached the downed Marine Huey and discovered a badly wounded crewman and four bodies. In his book *The War in I Corps*, Vietnam veteran Richard A. Guidry described the scene: "Outside the helicopter lay two dead NVA soldiers, their wounds still dripping blood. 'Five Marines, all dead,' someone from the search party called out from inside the helicopter. From the heap of bloody corpses an angry voice responded, 'I'm not dead, you idiots!'

"A badly wounded crewman lay beneath the bodies of our dead Marines. He told of how the helicopter was swarmed over by enemy soldiers who stripped it of everything they could carry away, including the watch from his wrist, as he played dead. More importantly, he said that one of the recon team had escaped into the brush." Bravo Company reported the information, evacuated the wounded crewman and started the search for Slocum. Before leaving, the company set fire to the wrecked helicopter.

By continuing to sneak about the battlefield, Slocum unknowingly presented himself as an enemy soldier. "Someone must have seen me and called in a fire mission," he recalled vividly, "about five times, only five or six rounds, but they didn't bother me because I took cover in one of the gook foxholes that covered the hill."

An aerial observer flew over to investigate the suspected enemy soldier. "He waved his wings," Slocum recalled, "and circled over me for about an hour before several Hueys showed up." The gunships strafed the ridge, trying to keep the NVA at bay. "The Hueys seemed to want me to move toward the grunts. I didn't want to do that because I didn't want to get shot again. I didn't have a weapon, and the gooks were between me and them." So Bravo Company moved toward the missing man. Slocum remembered, "The grunts started moving my way. First, I thought they were NVA, so I started moving the other way. The choppers sort of motioned me back in, and it was grunts after all. I walked over to them, and they had a medevac come in and pick me up." He was flown to Dong Ha for treatment, and after two and a half months recuperating returned to 3$^{rd}$ Force Reconnaissance Company.

The team's heroic and desperate fight for survival and rescue had finally ended.

**This story originally appeared in *Leatherneck* magazine in November of 2009. Dick Camp is a retired Marine colonel, and is indebted to LtCol George "Digger" O'Dell, USMC (Ret); Colonel Dave Underwood, USMC (Ret); Marine veteran Mike Nation; and former Hospital Corpsman Steve Thompson for their assistance.**

# STOLEN VALOR

*If the following series of stories ticks you off, that's okay - they were meant to.*

Rick Strandlof, executive director of the Colorado Veterans Alliance and the man most colleagues knew as Rick Duncan, was front and center during the 2008 political campaigns in Colorado. He spoke at a Barack Obama veterans rally in front of the Capitol in July, co-hosted several events with then- congressional candidate Jared Polis, and attacked Republican Senate candidate Bob Schaffer in a TV ad paid for by the national group Votevets.org - and the mostly Democratic candidates he supported, looking for credibility on veterans issues and the war, lapped it up appreciatively.

Now, politicians are dealing with news that the man they believed to be a former Marine and war veteran wounded in Iraq by a roadside bomb had in fact never served in the military - but *did* spend time in a mental hospital. Many of the candidates he supported won their elections handily, and now say they were defrauded as much as anyone else.

There is little doubt Strandlof had a remarkable ability to fool people, something aided by the fact that among his fabrications was a claim he suffered a severe brain injury, which helped cover behavior associates now concede was often erratic and strange - but there were also plenty of other signs during much of the time Strandlof was working on behalf of candidates for anyone watching carefully.

CVA wasn't registered as a political organization until well after the campaigns were over, and then only at the state level despite being active in federal campaigns, and although he claimed to represent thirty-two thousand veterans - the biggest post- 9/11 vets group in Colorado - Strandlof always showed up at events with the same small number of supporters. There were few concrete signs he represented more than the close circle he had gathered around him.

"Nobody really fully trusted any of those numbers. He had a few dozen people who were helping him out, and claimed to have a huge mailing list that no one ever saw. The VFW, the American Legion, none of those traditional veterans groups had ever heard of him," said one prominent veterans activist who worked for Democratic candidates during the campaign and spoke on condition of anonymity. "The veterans' community was very protective over him because he had portrayed himself as a wounded veteran. This is someone who claimed spending eighteen months undergoing physical rehabilitation after suffering debilitating injuries. You don't go at somebody like that hard. Perhaps we learned a good lesson here."

According to *The Associated Press*, CVA's board decided that the group would disband in the wake of news that Duncan was actually Strandlof.

*The following legal opinions just boggle my mind. Personally, I would like to see this lowlife locked in a room with a bunch of real wounded Marine Corps Silver Star recipients and let 'justice' take its course - but that's just me.*

Rick Strandlof may have lied about being a decorated Iraq War veteran, but those lies are protected by the First Amendment, according to his attorney and a civil liberties organization. Strandlof was charged in U.S. District Court in Denver with five misdemeanors related to violating the Stolen Valor Act - specifically, making false claims about receiving military decorations. He is accused of posing as "Rick Duncan," a wounded Marine captain who received a Purple Heart and a Silver Star. Strandlof used that persona to found the Colorado Veterans Alliance and solicit funds for the organization.

He was exposed when real veterans serving on the board of the Colorado Veterans Alliance became suspicious of Strandlof's claims, and the FBI opened an investigation. Then the Rutherford Institute, a nonprofit civil liberties group based in Virginia, filed a friend-of-the-court brief in Strandlof's case attacking the constitutionality of the Stolen Valor Act. John Whitehead, president of the Rutherford Institute, said the law is poorly written and should not be used to prosecute people for simply telling lies.

"You have to redraft the law to prove a particularized damage," he said. "If you run around Denver and yell out, 'I got the Medal of Honor,' you are guilty of the statute the way it is written."

In a recent motion to dismiss the case, Strandlof's attorney, Robert Pepin, wrote that "protecting the reputation of military decorations is insufficient to survive this exacting scrutiny," and Rutherford Institute attorney Douglas McKusick argued that Strandlof's lies did not "inflict harm" upon the medals he lied about or debase the meaning of the medals for the veterans who actually earned them. "Such expression remains within the presumptive protection afforded pure speech by the First Amendment," McKusick wrote. "As such, the Stolen Valor Act is an unconstitutional restraint on the freedom of speech."

The Stolen Valor Act prohibits people from falsely claiming they have been awarded military decorations and medals. The Act, signed into law in 2006, carries a punishment ranging from fines to six months in prison. Assistant U.S. Attorney Jeremy Sibert wrote that the false statements made by Strandlof are not protected speech. "Even if this court finds that the Stolen Valor Act affects protected speech and subjects it to the strict scrutiny standard of the First Amendment, the Act withstands the scrutiny because it serves the compelling interest of protecting the reputation and meaning of decorations and medals," Sibert wrote. "Since the Act's prohibition is narrowly tailored, its criminal penalty does not violate the First Amendment." Sibert also argued that punishing people for lying about military decorations does not chill the flow of political free speech or freedom of the press. "However, his lying about receiving military medals, false statements of fact in an effort to increase his status and credibility, fall into the class of unprotected 'utterances' that are not constitutionally protected."

# Been There, Done That... Got the T-Shirt!

*So, how does the non-military community feel about this case in particular, and the Stolen Valor Act in general? I offer this ridiculous editorial from the Denver Post as evidence that the protected masses have no clue about things related to service, sacrifice and entitlement:*

Rick Strandlof is accused of telling a lot of lies. Upon arresting him, federal authorities said he claimed to be a wounded Marine veteran who had received a Purple Heart and a Silver Star. Reprehensible? Yes. But criminal?

We have doubts about the constitutionality of the 2006 Stolen Valor Act, which makes it a crime to merely say you had received certain military decorations when you hadn't, because the First Amendment protects even deplorable, distasteful speech - particularly in cases where that speech doesn't injure someone else.

Strandlof's deception began to unravel when members of the group he founded, the Colorado Veterans Alliance, began checking out his claims of having served three tours in Iraq, surviving the 9/11 attacks on the Pentagon and suffering a brain injury from an improvised explosive device.

It has not been shown Strandlof lied to gain anything for himself other than publicity. He is accused of using his created persona to solicit donations on behalf of veterans. We are not defending the lies, but there hasn't been a case made he was making such claims to line his own pockets.

You might argue that the "injured party" is really the integrity and honor associated with the awards - that they are cheapened by frauds who falsely claim them. Frankly, if anyone is injured by such false claims, it would be the person who is lying. The veterans community that Strandlof had so integrated himself into turned on him, and for good reason, once they realized he wasn't who he said he was.

The Stolen Valor Act, introduced in the U.S. House by Colorado's Rep. John Salazar, broadened the provisions of prior U.S. law that prohibited the unauthorized wearing or manufacturing of military decorations and medals. Salazar's website says the congressman supported the measure because "These imposters degrade the meaning of medals earned in service to our nation and sometimes use their 'standing' as a medal recipient to commit further fraud and more dangerous crimes."

Pursuing fraud charges against those who make false claims to enrich themselves or hurt others is one thing. But criminalizing the mere act of lying is entirely another. Would truth police squads pursue all lies? There is a better way to protect the integrity of military awards. First, make the names of all recipients of military decorations widely available. Some are not. That way, it would be easy to check someone out.

The question of whether the Stolen Valor Act is constitutional will play out in the court system. We hope the judicial system will recognize the value of free speech, even when it's not popular.

*Did this moron REALLY just say he thinks this Strandlof character is the REAL victim, and go on to suggest this sort of "free speech" has some sort of value?*

# COUNTERFEIT HERO

*It is one thing to be a "poser" and lie about your service, and another altogether to lie about holding the Medal of Honor. This appeared on Thursday, January 17, 2008 in the Inland Valley (California) Daily Bulletin:*

Xavier Alvarez is a cretinous, low level elected official out of Pomona, California, specifically the District I Representative on the Three Valleys Water Board. This low-life appears to be more of a political gadfly than a civic minded citizen, having run for other local offices and usually finishing dead last, so he thought it might help his chances if he padded his résumé a bit. Now, it's one thing to lie about something - but to lie about something as easily verifiable as this is idiotic! If you go to the website cmohs.org you will find the names of the living Medal of Honor recipients - and there is no one by that name listed. As a matter of fact, you will not find the name Alvarez listed anywhere as being a recipient of the Medal of Honor.

In his biography for the Three Valleys Water Board, he says he is a "member of American Legion Post 1000 based in San Francisco, and a member of American Veterans." First of all, there is no local post 1000 in San Francisco. Post 1000 is the state headquarters. If he is a member of the American Legion, then he presented a falsified DD-214. That is a felony.

In any case after Mr. Alvarez made his claims, someone

did check a list and found out this butt nugget huckster was lying. He was reported to the feds, and the feds have since charged him with violating the Stolen Valor Act. But wait! Alvarez now admits his fantasy, but says it's his right under the Constitution to make such claims and that the Stolen Valor Act is unconstitutional because it takes away his right of free speech.

There is actually a link on the internet where one can hear this jerk call himself a wounded, retired Marine and a recipient of the Medal of Honor. He was recorded making the original Medal of Honor claim in July after he was introduced as a guest at a Walnut Valley Water District board meeting, saying, "I'm a retired Marine of twenty-five years. I just retired in 2001. Back in 1987 I was awarded the Congressional Medal of Honor. I got wounded many times. At the same time, I'm still around." It's nauseating!

This is Alvarez's defense - after admitting he never served in the military, he was charged with violating the Stolen Valor Act and pled not guilty to the misdemeanor charge. Alvarez's lawyer, Brianna J. Fuller, argued in the motion to dismiss that "protecting the reputation of military decorations" is not a compelling enough reason to place "restrictions on false statements."

Yeah, you read it right. In other words, military heroism is just not worth protecting. Who thinks like that? When he made the claim he damned well thought it was worth something or he wouldn't have made it, but now that he's been ratted out as a liar and the piece of vomit that he is it suddenly doesn't mean anything.

I doubt a judge will buy that idiotic reasoning (except that this is California we are speaking of, so there is a chance) but unfortunately Alvarez will probably get off with a hand slap, when what he deserves is to be put in a room full of Marines and let them have fifteen minutes with him. Maybe they could introduce him to a *real* Water Board!

**The Stolen Valor Act, under which Alvarez was fined and ordered to perform community service in 2007, was upheld as being unconstitutional by the 9th U.S. Circuit Court of Appeals, which upheld an earlier ruling determining that a law barring people from lying about their military heroics was a violation of free speech. The earlier ruling, which was made by three of the court's members, invalidated the 2006 Act by Congress. Alvarez pled guilty in July 2008 to falsely saying he had been awarded the Medal of Honor, was fined $5,000, and sentenced to three years of probation which required community service. He is currently at Centinela State Prison in Imperial County for defrauding the water district after being convicted of registering an ex-wife for health benefits with the district in 2007. His public defender is expecting the case to be appealed to the U.S. Supreme Court.**

# THE COURTS HAVE STOLEN
## *The Stolen Valor Act*

Thomas D. Segel

Those who have worn the Dress Blue uniform of the United States Marine Corps know the red stripe running down the trouser legs represents blood which has been shed for this country. They also know the various medals for valor or military merit are never "won," but are instead awarded in recognition of specific acts by an individual member of the armed forces.

During my decades of Marine Corps service plus a journalism career I have been very fortunate to have known dozens upon dozens of our nation's heroes. In that distinguished band of brothers were men who were awarded the Medal of Honor, the Navy Cross and multiple awards of Silver and Bronze Stars for exceptional valor. Also among that proud assembly should be listed those almost uncountable numbers of men and women who have received the Purple Heart Medal for wounds inflected by the enemy. I still feel that I was privileged to stand in the shadows of such worthy members of the warrior clan.

While writing about everything from combat action to individual exploits across the years, I have also periodically encountered those vile pretenders who falsely portrayed themselves as soldiers, sailors, airmen or Marines. These frauds generally advanced stories that they had a history of performing heroic acts and were highly decorated by a grateful nation. Some of these military fakes were quickly unmasked, but there were others who advanced themselves through the stolen valor of real heroes for a number of years.

Finally, through the efforts of some dedicated veterans, the 2005 Stolen Valor Act was passed by Congress and signed into law - making it a crime punishable by up to a year in prison for falsely claiming to have received high military decorations.

Now the liberal leaning 9th U.S. Circuit Court of Appeals has ruled the Stolen Valor Act is an unconstitutional restraint on free speech and a threat to every citizen. Said Chief Judge Alex Kozinski when rendering the decision of the court: "Saints may always tell the truth, but for mortals, living means lying."

Judge Diarmuid F. O'Scannlain, writing in dissent, claims the majority has misinterpreted forty years of Supreme Court decisions and added the high court has consistently held that "the right to lie is not a fundamental right under the Constitution."

Using this court decision as a foundation, I would assume the 9th Circuit would also declare laws against lying to federal officers, IRS agents or Congressional Committees are also unconstitutional. That would free many citizens who are now looking at the world from behind the barred windows of prisons. After all, the Court said for mortals, living means lying. It should also be pointed out that Chief Judge Kozinski feels the Stolen Valor Act "is a threat to every citizen who fibs to embellish his or her image."

Unless this despicable decision is referred on to the Supreme Court and reversed, the country can again expect phony heroes to creep out from under their respective rocks across the United States. After all, why should they be punished for pretending to be valiant? Members of Congress and the Executive branch of government lie to Americans every day - and they too pretend to be heroes.

**Thomas D. Segel is a twice-wounded former combat correspondent who retired after twenty-six years of service in the Marine Corps. He holds eight personal decorations for valor and meritorious service, received the Thomas Jefferson Award for Journalistic Excellence, has been named Military Writer of the Year, and is a past national president of the Marine Corps Combat Correspondents Association. His work can be found at www.thomasdsegel.com**

# HARVEST MOON

### Colonel Harvey Barnum

*After making readers suffer through the tale of a bogus MOH recipient, I thought it only fair to balance the books and follow with the story of a bonafide hero. We all salute you, sir!*

I was stationed in Okinawa with the 3rd Marine Division in 1963 as the fire-support coordinator for the 9th Marine Infantry Regiment. As you recall, we mounted out on three or four different occasions that year to go to Vietnam to secure the hills around the Dan Nang air base. We never got there.

On one occasion I was in the Philippines, firing naval gunfire, and was called back in from the naval gunfire range to Subic bay. The ships were going to pick up my forward observers and naval gunfire spotters on the way to Vietnam. That never materialized.

I was aware of what was going on, because I was in on the preparation. On two other occasions we boarded aircraft, took off, and returned to secure Kadena Air Base on Okinawa. The missions changed from contingency movements to exercises when entry into Vietnam was called off.

I was stationed at Marine Barracks, Pearl Harbor, when one of the officers at Marine Barracks, Barber's Point, a

249

Captain Gardner said, "You know, it's kind of tough having these young troops here in the Pacific guarding doors, classified information and buildings while there is a war going on."

After commenting about this situation to his fellow officers, Captain Gardner drafted a proposal suggesting an alternative. He proposed sending company-grade officers and staff NCOs to combat in Vietnam for a two-month period to serve in their individual MOS (military specialty), and come back to the barracks in the Pacific to tell the troops what was going on over there. He felt this would help motivate the troops by showing them what they were doing in Hawaii was very important even though there was a conflict going on at the time in Southeast Asia.

The program was approved by General Victor Krulak, Commanding General Fleet Marine Force (Pacific), and Marine Barracks, Pearl Harbor was assigned the first quota - an officer and a staff NCO. At that time there were, I believe, twenty-four officers in the barracks, and I was the only bachelor. I had already augmented, indicating that I was planning to stick around the Marine Corps for a while. Consequently, I volunteered to go do what I had come into the Marine Corps to do.

"If this is going to be my chosen way of life," I reasoned, "then I'd better find out what it's all about." Of course in addition to career considerations, there was a great deal of adventure, excitement, anticipation - launching off into the unknown.

I went to my commanding officer and said, "You know,

it's going to be over the holiday season, Christmas and New Year's, so I feel I should be the one to volunteer. Why send a married man? He'll get his opportunity in the months to come." Boy, was that a ploy to get that assignment, but it worked.

My CO approved my request, and myself and First Sergeant John Matson - a big man from security company - were the first to go to Vietnam on the program. I was assigned to Echo Battery, 2nd Battalion, 12th Marines - the same battery that I commanded later during my '68-'69 tour.

The purpose of the program, in my particular instance, was to allow me the opportunity to serve as a forward observer, fire-direction officer and XO, and then come back and relate my experiences to the troops - but it was more than that. Vietnam, I mean. There was another side that people back home didn't hear about.

We were involved in the military side of the war, of course, but we also participated in the people-to-people aspect of the war. We did a lot of work with the people and the local orphanages, and we worked hard. These activities didn't receive much press, but it was still something we all were proud of.

After my (Medal of Honor) action, I also served with Lima Battery, Four-Twelve, for a couple of weeks to get a feel for the '55s. There were major differences between the '05s and the '55s and it was vital to understand them with regard to the security aspects for movements, employment and deployment of the towed, self-propelled guns.

When I got in-country, the first thing I noticed about

Vietnam was the smell of latrines burning at Da Nang. It was really something. We used fifty-five gallon drums cut in half and they were burned off every day - a smell no Vietnam vet will ever forget.

Another thing was that when I got there in December 1965, there had been a lot of rain and the red mud was knee-deep. This was before there were many roads in Da Nang, before the buildup.

When I arrived, I joined my unit south of Da Nang. We were located near the CP of 2nd Battalion, 9th Marines in a stationary position firing in support of Two-Nine when they went out on their daily patrols. I was assigned as the forward observer for Hotel Company and I joined them in their position on the battalion perimeter. We were on patrol on the Anderson Trail, where one of the first big ambushes took place - VC ambushed Marines - and we were recalled. If I remember, this happened on my second patrol with the unit.

We immediately terminated the patrol and were airlifted out to replace Fox Company from Two-Seven. They had been in a firefight and had several battlefield casualties, as well as a great number of immersion foot casualties. They had been participating in Operation Harvest Moon for several days. Hotel Company joined Two-Seven for the remainder of the operation.

Around noon on the 18th of December, the lead elements of the battalion march column had entered into Ky Phu and had proceeded on through. H&S Company was just entering the village, and we heard firing.

Hotel Company was the real element, tail security, and I

was thirty or thirty-five yards behind Hotel company commander Captain Paul Gormley. His radioman was right behind him. We were coming around this little hill with rice paddies to the right and about two or three hundred meters between our lead element and the village of Ky Phu.

I still say the initial round that triggered the ambush on the rear element hit the company commander. I'm pretty sure it was a rocket, probably a B-40 rocket, and then all hell broke loose. It happened that quick.

It was a matter of minutes. I had heard firing in the distance, up toward Ky Phu, and all of a sudden we were right in the middle of it. The commander was walking along with the radio antenna sticking up and they picked it out. And, boy, they did it right. They took him out.

Of course, we all took cover and started returning fire. I remember the corpsman running by me - Wesley "Doc Wes" Berrard, a black corpsman from the Chicago area if I recall. He called back and said, "The company commander is seriously wounded." Then Doc Wes was shot. My scout sergeant ran forward to help the Doc and cover him with return fire. Then he was shot.

At that point I ran forward to the company commander and his radio operator. The young radio operator had been killed outright, and Captain Gormley was mortally wounded. I picked him up, carried him back to safety, and he died in my arms. Then I went back out and carried Doc Wes back.

I returned to where Captain Gormley and the radio operator had been hit and took the radio off the dead operator, a PRC-25, and strapped it to myself. I called in

253

artillery, but we were at max range. Jerry Black was battery commander, and boy, he was giving it all he had! We also had helicopter gunfire support.

It wasn't a VC ambush - it was NVA. I got close to some of the NVA and fired my .45 pistol a couple of times. I also had an M14 rifle and went through my ammunition rather rapidly, because you could see where the enemy fire was coming from across the rice paddy less than fifty meters away. As I used up my ammunition, I went from the firing mode into the listening, analyzing, and giving direction mode.

I got on the radio and told our battalion commander, Lieutenant Colonel Leon Utter, what had happened, the condition we were in and that I had assumed command of the company. I don't think there was an XO, I don't remember it's been so long, but I was not the senior guy. One of the platoon commanders had seniority and could have assumed command of the company, but I was there and I had things in motion. I told Colonel Utter, "The platoon commanders have their hands full. I am aware of what is going on and I have assumed command."

He told me, "Continue to march and make sure everyone knows you are the boss." I did.

I think one of the things that was phenomenal about the battle is for the next three or four hours all company commanders, the battalion commander and the helicopters were on the same radio frequency. By listening, you could tell what was going on, and when you figured that you were in a worse situation than the other guy, then you would butt in on the frequency.

I led a couple of counter-attacks against the enemy positions to our right flank and rear. There was incoming fire off to our right flank, and I went up on this little hill to get a better view. I could see where the fire was coming from and I pointed out targets for our Huey gunships to take under fire.

I knew that I would have better luck with the gunships than the artillery, because the enemy was entrenched at the edge of the artillery's maximum range. In order for the artillery to bring fire to bear on them they would have had to fire directly over our heads - we were right on the gun-target line. At that range the chance of error was significantly greater than closer in. The Hueys were there, so I used them.

I pointed out targets for the choppers and I remember firing a Willy Peter (white phosphorus) rocket as a marking round once. Hell, those Hueys were coming in right over our heads and I was standing up pointing at the damn targets and talking to the pilots.

I would give them a target heading, and when they could pick me out visually, pointing with my arm to the target, they would come in. By that time, though, they were getting shot at too, so they could see where the fire was coming from. At the same time I was directing one of my platoons in a successful counterattack against the NVA positions on our right flank.

We were getting pretty low on ammunition and we suffered a number of casualties. It was starting to get dark - it was also an overcast day, if I recall - so we cleared an area of some trees to bring in the H-34 choppers to take out the wounded and dead.

The battalion commander said, "You've got to come out and join up with the remainder of the battalion, and you've got to come out by yourself." The rest of the battalion was having a hell of a battle in the village and could not come to our aid. So I told everyone to lighten their load, to drop their packs in a pile, and any equipment that was not working, including radios and machineguns, and I had the engineers blow it up.

The Marines in the village set down a base of fire and we commenced squad rushes across three hundred yards of open fire-swept ground. If someone fell, someone else picked him up, and we brought everybody out. It was really something to see. Teamwork at its best!

That was about four and a half hours of battle. I did what I had been trained to do. I made decisions and people carried them out. That was the most amazing thing. Here I was company commander, and most people didn't even know who I was. I was simply the officer who stepped forward and took command. Despite being relatively unknown, people did what I told them to do, when I told them, and in the manner I told them to do it. Some of them got hurt, some of them got killed, but they still carried out their orders. The result was success.

When we rejoined the battalion, I met with the battalion commander and he assigned Hotel Company a sector of the perimeter to defend. I set my people out on the perimeter and sat down with the gunny to account for the dead and wounded and make arrangements for resupply.

They mortared us that night, pretty much throughout the night. The next morning I remember crawling on my belly with General Jonas Platt, the task-force commander, and surveying the battlefield. Of course, they had removed most of the bodies by that time. The rest of that day we maneuvered to Route 1. Two-Seven mounted up and went south to Chu Lai and we got on trucks and went north to Da Nang to rejoin Two-Nine.

When we started loading the trucks, we took small-arms fire from a nearby village. We had just come through one hell of a battle, and these Marines were some pretty seasoned guys. After you come through what we had come through, you don't know what incoming rounds might turn into. We were grouped up to load the trucks and it was a dangerous time to get bunched up.

An Ontos (a small, tracked vehicle armed with six 106-mm recoilless rifles) had come down the road to meet us for road security, and I told the Ontos crew to turn around and fire on the hut from where we had received that incoming fire. The Ontos fired, leveled the hut, and the fire ceased. Someone later criticized me for this, saying that I had over-reacted. Well, I had seen that one round could signal the start of a hell of a fight - therefore I made the decision to eliminate that hut as a threat. It was eliminated, and I stand by my decision.

Anyway, the Battle of Ky Phu and Harvest Moon was over, and they figured we overcame twelve-to-one odds. We got on the trucks and went north and stayed the night at

FSLG (Force Service Logistics Group) between Da Nang and Chu Lai. While there I was awakened by a colonel from Division HQ around midnight, who took me to a tent and started asking questions about the battle. He said, "You're probably going to get a Sunday School Award out of this."

The next morning a new company commander came in and I turned Hotel Company over to him. When I got back to Da Nang my feet were in pretty bad shape from immersion foot, and I was taken over to the hospital and had the water blisters drained. Then I went over and got cleaned up - we hadn't showered or shaved for five or six days - and I went to bed.

I wasn't sent out on any more patrols and stayed in the battery area until I went over to Lima, Four-Twelve, to work with that battery. One night the battery commander woke me up and said, "Boy, you've got to get up. General Walt (Commanding General, 3rd Marine Amphibious Force) wants to see you." That's when he told me General Walt had recommended me for the Medal of Honor.

I can remember waiting to see him and being scared. I thought, "I'm going to meet a three-star general." I don't think I had ever met a three-star general, much less the MAF commander. He talked to me like a son, though. I realized very quickly that there was a lot of compassion, a lot of concern in the general - but I remember how those blue eyes of his penetrated right through me.

Later, when he came in to speak to the 9th Marines, I remember he jumped up on the hood of a jeep and commented on my action. I guess I was his teaching point.

People treated me with a great deal of respect, because they knew I'd been recommended for the Medal of Honor. I recall that on Christmas I was even able to attend Cardinal Spellman's Mass.

I wanted to stay in Vietnam for a full tour, but left in February at the termination of my temporary duty. When I returned to Marine barracks in Hawaii, two more officers and two more NCOs from Security Forces (Pacific) went on this program.

I was at Fort Sill in the summer of '67, and went to the Marine Corps League Convention in St. Louis. General Walt was there, and one of his aides told me the general was looking for another aide and asked if I was interested. I said, "Well, I've got orders for the 2nd Division."

From St. Louis I flew to New Orleans, where I got the Military Man of the Year Award and General Walt presented it. After he hung it around my neck he said, "I want to see you afterwards." I went to see him and he said, "Do you want to be my aide?"

I said, "I'd be honored, but I've got orders for the 2nd Marine Division."

He said, "Go home and you'll receive orders when you get there to report to Headquarters Marine Corps." When I got home, I had orders assigning me as his administrative assistant. I was his aide for a year. At the time General Walt had said he didn't think an aide should be an aide for more than a year.

Now, when you complete an aide assignment and have done a good job, it's pretty much understood that you will get the next assignment you want - so, after I completed my assignment as his aide I said, "I want to go back to Vietnam." Keep in mind there was a little pressure at the time about me not going back into combat. He felt the way I did - that Vietnam was the place to be - where Marines should be. Here I am wearing the Medal of Honor, and I hadn't even served a full tour of duty in Vietnam. General Walt was a man of his word.

Going back, there's a week refresher course at Camp Pendleton and, unbeknownst to me, General Walt had contacted General Ray Davis - Medal of Honor, Korea - who had the division, and they had a battery all set for me when I got there. I got to Camp Pendleton and checked in at four o'clock one morning, and a young lance corporal said, "Sir, the next company grade officer coming in is supposed to take C Company through staging battalion." I replied, "If these are your orders, I am ready to carry them out."

I stayed at staging battalion for a month and got to Vietnam six weeks later. All this time that battery assignment awaited me, and when I finally showed up General Davis said, "Where have you been?"

I'm glad I took that company through staging, though. Not only did I help prepare the young Marines for combat in Vietnam, but it also got me prepared physically and mentally. I took over Echo Two-Twelve, my old unit.

We were fire basing all the time - we built or reoccupied over twenty fire bases. I've got a diary at home with all the details. We saw a good deal of action. We went through the A Shau Valley with the Ninth Marines - Operation Dewey

Canyon. We were firing in support of Wes Fox's Alpha One-Nine during his Medal of Honor action.

I did, and my Marines did, what we were sent there to do. We followed orders, and considering the restraints under which we were operating, I think we did damn well. We carry on our shoulders a proud tradition that has been molded by hundreds of thousands who have gone before us, and I'll be damned if we are going to let them down. There have been several times when the odds were phenomenal and how did Marines come through it? Semper Fi.

Marines don't *say* that they can do the job. They *do* it. That's why Americans believe in us. I swear to God. That is one thing I always remembered. If I got shot, a Marine would never leave me on the battlefield. Marines got killed going to the aid of other Marines. That is something we know, and we are proud of it. And *lead!* If you can't stand up and lead, then get the hell out of the way. Someone else will do it. Don't tell your people with your hands on your hips to run around the grinder. You be up front running, and let them follow you.

Since I've been in Washington people come to visit and I take them to The Wall, and it affects me differently every time. The first time I went was during the Inauguration several years ago. Wheew! That was a tough day, really tough. That visit shot the rest of the day. I couldn't go out. The gal I was with couldn't believe the effect it had on me.

First of all, I say this - the location of it overpowers the design - the Lincoln Memorial on one hand, and the Washington Monument on the other. What a place of honor.

There is not just another war memorial in Washington with that prominence. I'm proud of it, and what it represents.

In November a few years ago I was there with my mom and dad, and I got a little misty eyed. Sometimes, emotionally, I just go to pieces, privately. Then a couple of weeks later I was there, and I walked away with a little bounce in my step. I really felt good.

**Courtesy of *Valor* by Timothy S. Lowry**

# LAST DAYS FOR MARINE
## *Were True Finest Hours*

**Denis Hamill**

*As a Marine, New Yorker, and old guy this gives me hope there are still people out there who know how to do the right thing. Officer Porcello was nominated and became a finalist for the "America's Most Wanted All Star Award," but somehow, even though I voted for her (numerous times), she didn't win. No matter - she has received numerous other honors, and will always be a winner in my book.*

Sometimes when old Marines die they do fade away into unmarked graves in Potter's Field.

Such might have been the case for Gaspar Musso, USMC 925050, who fought in the Battle of Tinian in the Marianas Islands in 1944 and who died on November 15, 2008 at age eighty-four in a Brooklyn nursing home.

Enter Police Officer Susan Porcello, a PBA delegate at the 68th Precinct in Bay Ridge and one of those big-hearted New Yorkers who still make this the best city on Earth.

"No way was I going to let this brave old Marine who fought for his country in WWII get buried in Potter's Field," she says.

Porcello first met Musso back in July when she responded to a 911 ambulance call to the retired insurance broker's one-bedroom apartment on, appropriately, Marine Avenue.

"When my partner, Eddie Ennis, and I arrived at his apartment Gaspar seemed a little bit down about himself," Porcello says. "He said he felt alone in the world. We talked to him a bit and as I looked around his tidy apartment I noticed that he had served in the military - the Marines to be exact."

Porcello asked him about family and friends. "Look around you, what do you see?" Musso said. "I have no family or friends."

To which Porcello said, "Well, I'm your friend."

Right there, with those four beautiful words, Gaspar Musso was destined to die with the dignity he'd earned with a rifle in his hands, fighting in a USMC uniform, in a war that saved civilization.

If she didn't already wear a badge, you'd want to pin a star on Susan Porcello.

Musso, a diabetic with a host of other age-related maladies, had accidentally overdosed on his prescription medications and Porcello accompanied him to Lutheran Medical Center.

"I told him I'd be back to visit him and take him to a senior center where he could make some friends," said Porcello, who comes from a big Italian family with a mom, dad, three sisters and a brother. "I told him I was making him my 'Grandpa,' and if he liked, he could spend Thanksgiving with my family. Eddie and I discussed alternating holidays with Gaspar so he wouldn't be alone for any of them."

Two days later Musso was placed in critical care. Porcello asked hospital staff where he'd be buried if he didn't make it. "Potter's Field," said one administrator.

"This infuriated me," said Porcello. "There was no way I was going to let a man who fought for our country be buried in Potter's Field. Not on my watch!"

Porcello told the hospital to keep her apprised of Musso's condition. She had a local priest visit him. Porcello even asked NYPD's Missing Person's Squad to search for next of kin - but with no luck. Musso had been an only child to Anthony and Marie Musso, both deceased. He had no other relatives. Musso's only friend, an upstairs neighbor, had died the year before.

After his health improved, Musso asked Porcello to become his official health proxy. She transferred him to Caton Park Nursing Home, where he was treated extremely well. She visited him often, learning that Musso was born May 7, 1924, had joined the USMC in December of 1943, finished training at Camp Lejeune in March 1944 and was fighting with the 2nd Marines on Tinian Island by July 1944.

"I visited Gaspar on November 13, bringing him rosary beads, a Bible, and his reading glasses," she said. "The next day I returned and found Gaspar sitting up in a chair, dressed in his own clothes. Looking great." Porcello washed his hands and face, trimmed his nails and eyebrows and asked if he was coming to her house for Thanksgiving. "I'm trying!" he said. He also asked Porcello to bring him a Christmas wreath for his room.

The next morning Porcello received a phone call saying that Gaspar Musso had died peacefully in his sleep. No way was she going to let her good friend be toe-tagged and buried in Potter's Field. Porcello paid out of her own pocket for a wake at McLaughlin's on Third Avenue and a Mass at St. Patrick's Church in Bay Ridge, where a crowd of good-hearted cops from the 68th Precinct filled the pews, with six serving as pallbearers. Sergeant Angel Rosa of the 68th, also a Marine, arranged for a USMC honor guard at Musso's funeral.

Then taps blew over Gaspar Musso, United States Marine, as he was buried next to his mother at Resurrection Cemetery in Staten Island. With the dignity he deserved.

Semper Fi, Gaspar.

**This story originally appeared in the *New York Daily News* on December 09, 2008. On February 10, 2009 the New York State Legislature passed a Resolution honoring Officer Porcello, and on February 17 the first annual "Susan Porcello Day" was held at the Fort Hamilton Senior Center.**

# THE DOG

They told me the big black Lab's name was Reggie as I looked at him lying in his pen. The shelter was clean, no-kill, and the people really friendly. I'd only been in the area for six months, but everywhere I went in the small college town, people were welcoming and open. Everyone waved when you pass them on the street.

But something was still missing as I attempted to settle in to my new life here, and I thought a dog couldn't hurt. Give me someone to talk to. And I had just seen Reggie's advertisement on the local news. The shelter said they had received numerous calls right after, but they said the people who had come down to see him just didn't look like "Lab people," whatever that meant. They must've thought I did.

At first I thought the shelter had misjudged me in giving me Reggie and his things, which consisted of a dog pad, bag of toys - almost all of which were brand new tennis balls - his dishes, and a sealed letter from his previous owner.

You see, Reggie and I didn't really hit it off when we got home. We struggled for two weeks (which is how long the shelter told me to give him to adjust to his new home). Maybe it was the fact that I was trying to adjust, too. Maybe we were too much alike.

For some reason his stuff (except for the tennis balls - he wouldn't go anywhere without two stuffed in his mouth) got tossed in with all of my other unpacked boxes. I guess I didn't really think he'd need all his old stuff, that I'd get him new things once he settled in. but it became pretty clear pretty soon that he wasn't going to.

I tried the normal commands the shelter told me he knew, ones like "sit" and "stay" and "come" and "heel," and he'd follow them - when he felt like it. He never really seemed to listen when I called his name - sure, he'd look in my direction after the fourth of fifth time I said it, but then he'd just go back to doing whatever. When I'd ask again, you could almost see him sigh and then grudgingly obey.

This just wasn't going to work. He chewed a couple of shoes and some unpacked boxes. I was a little too stern with him, and he resented it. I could tell. The friction got so bad that I couldn't wait for the two weeks to be up, and when it was I was in full-on search mode for my cell phone amid all of my unpacked stuff. I remembered leaving it on the stack of boxes for the guest room, but I also mumbled, rather cynically, that the "dog probably hid it from me."

Finally I found it, but before I could punch up the shelter's number I also found his pad and other toys from the shelter. I tossed the pad in Reggie's direction and he sniffed it and wagged his tail with the most enthusiasm I'd seen since bringing him home - but then I called, "Hey Reggie, you like that? Come here and I'll give you a treat." Instead, he sort of glanced in my direction - maybe "glared" is more accurate - and then gave a discontented sigh and flopped down. With his back to me.

Well, that's not going to do it either, I thought, and I punched the shelter phone number - but I hung up when I saw the sealed envelope. I had completely forgotten about that, too. "Okay, Reggie," I said out loud, "let's see if your previous owner has any advice..."

The letter began, "To Whomever Gets My Dog - Well, I can't say that I'm happy you're reading this, a letter I told the shelter could only be opened by Reggie's new owner. I'm not even happy writing it. If you're reading this, it means I just got back from my last car ride with my Lab after dropping him off at the shelter. He knew something was different. I have packed up his pad and toys before and set them by the back door before a trip, but this time... it's like he knew something was wrong. And something is wrong... which is why I have to go to try to make it right.

So let me tell you about my Lab in the hopes that it will help you bond with him and he with you.

First, he loves tennis balls. The more the merrier. Sometimes I think he's part squirrel, the way he hordes them. He usually always has two in his mouth, and he tries to get a third in there. Hasn't done it yet. Doesn't matter where you throw them, he'll bound after it, so be careful - really don't do it by any roads. I made that mistake once, and it almost cost him dearly.

Next, commands. Maybe the shelter staff already told you, but I'll go over them again. Reggie knows the obvious ones – "sit," "stay," "come," "heel." He knows hand signals. "Back" to turn around and go back when you put your hand straight up, and "over" if you put your hand out right or left. "Shake" for shaking water off, and "paw" for a high-five. He does "down" when he feels like lying down - I bet you could work on that with him some more. He knows "ball" and "food" and "bone" and "treat" like nobody's business.

I trained Reggie with small food treats. Nothing opens his ears like little pieces of hot dog.

Feeding schedule is twice a day, once about seven in the morning, and again at six in the evening. Regular store-bought stuff. The shelter has the brand.

He's up on his shots. Call the clinic on 9th Street and update his info with yours. They'll make sure to send you reminders for when he's due. Be forewarned - Reggie hates the vet. Good luck getting him in the car - I don't know how he knows when it's time to go to the vet, but he knows.

Finally, give him some time. I've never been married, so it's only been Reggie and me for his whole life. He's gone everywhere with me, so please include him on your daily car rides if you can. He sits well in the backseat, and he doesn't bark or complain. He just loves to be around people, and me most especially.

Which means that this transition is going to be hard, with him going to live with someone new - and that's why I need to share one more bit of info with you... his name's not Reggie.

I don't know what made me do it, but when I dropped him off at the shelter I told them his name was Reggie. He's a smart dog, he'll get used to it and will respond to it, of that I have no doubt... but I just couldn't bear to give them his real name. For me to do that, it seemed so final, that handing him over to the shelter was as good as me admitting that I'd never see him again. And if I end up coming back, getting him, and tearing up this letter, it means everything's fine. But if someone else is reading it, well... well it means that his new owner should know his real name. It'll help you bond with him. Who knows, maybe you'll even notice a change in his demeanor if he's been giving you problems.

His real name is Tank... because that's what I drive.

Again, if you're reading this and you're from the area, maybe my name has been on the news. I told the shelter that they couldn't make "Reggie" available for adoption until they received word from my company commander. See, my parents are gone, and I have no siblings, no one I could've left Tank with... and it was my only real request of the Corps upon my deployment to Iraq, that they make one phone call to the shelter... in the "event"... to tell them that Tank could be put up for adoption. Luckily my colonel is a dog guy too, and he knew where my platoon was headed. He said he'd do it personally. And if you're reading this, then he made good on his word.

Well this letter is getting to be downright depressing, even though, frankly, I'm just writing it for my dog. I couldn't imagine if I was writing it for a wife and kids and family. But still, Tank has been my family for the last six years, almost as long as the Marine Corps has been my family - and now I hope and pray that you make him part of your family and that he will adjust and come to love you the same way he loved me.

That unconditional love from a dog is what I took with me to Iraq as an inspiration to do something selfless, to protect innocent people from those who would do terrible things... and to keep those terrible people from coming over here. If I had to give up Tank in order to do it, I am glad to have done so. He was my example of service and of love. I hope I honored him by my service to my country and comrades.

Alright, that's enough. I deploy this evening and have to drop this letter off at the shelter. I don't think I'll say another good-bye to Tank, though. I cried too much the first time. Maybe I'll peek in on him and see if he finally got that third tennis ball in his mouth.

Good luck with Tank. Give him a good home, and give him an extra kiss goodnight - every night - from me. Thank you, (signed) Paul Mallory.

I folded the letter and slipped it back in the envelope. Sure I had heard of Paul Mallory. Everyone in town knew him, even new people like me. Local kid, killed in Iraq a few months ago and posthumously earning the Silver Star when he gave his life to save three buddies. Flags had been at half-mast all summer.

I leaned forward in my chair and rested my elbows on my knees, staring at the dog.

"Hey, Tank," I said quietly.

The dog's head whipped up, his ears cocked and his eyes brightened. "C'mere boy."

He was instantly on his feet, his nails clicking on the hardwood floor. He sat in front of me, his head tilted, searching for the name he hadn't heard in months.

"Tank," I whispered.

His tail swished.

I kept whispering his name, over and over, and each time, his ears lowered, his eyes softened, and his posture relaxed as a wave of contentment just seemed to flood him. I stroked his ears, rubbed his shoulders, buried my face into his scruff and hugged him.

"It's me now, Tank, just you and me. Your old pal gave you to me." Tank reached up and licked my cheek. "So whatdaya say we play some ball? His ears perked again. "Yeah? Ball? You like that? Ball?" Tank tore himself from my grasp and disappeared into the next room - and when he came back, he had three tennis balls in his mouth...

# MRE DINNER DATE

*The following is a true story, and I was in tears reading it. For those of you who know what an MRE is, this is absolutely hilarious... and I can definitely understand how the young woman felt. I kept thinking of all the concoctions I used to come up with in the field to make an MRE edible, and as I recall the first thing in my rucksack was always a bottle of hot sauce. I thought I had made every concoction possible until I read this story.*

I had a date the other night at my place, and on the phone the day before the girl had asked me to "Cook me something I've never had before" for dinner.

After many minutes of scratching my head over what to make, I finally settled on something she definitely, *definitely* had never eaten before. I got out my trusty case of MRE's. (Meal, Ready-to-Eat) Field rations, which when eaten in their entirety contain three thousand plus calories each.

Here's what I made. I took three of the Ham Slices out of their plastic packets, took out three of the Pork Chops, three packets of Chicken-a-la-king and eight packets of dehydrated butter noodles and some dehydrated/rehydrated rice. I cooked the Ham Slices and Pork Chops in one pan, sautéed in shaved garlic and olive oil. In another pot, I blended the Chicken a-la-king, noodles and rice together to make a sort of mush that looked suspiciously like succotash. I added some spices, and blended everything together in a glass pan which I then cooked in the oven for about thirty-five minutes at 450 degrees.

When I took it out it looked like, well, ham slices, pork chops and a bed of yellow poop. I covered the tops of the meat in the MRE cheese (kinda like Velveeta) and added some green sprinkly things from one of my spice cans (hey, if it has green sprinkly things on it, it looks fancy right?) For dessert I took four MRE Pound Cakes, mashed 'em up, added five packets of cocoa powder, powdered coffee cream, and some water. I heated it up and stirred it until it looked like a sort of chunky gelatinous poop, and I sprinkled powdered sugar on top of it. Voila! Anger Pudding. For alcoholic drinks, I took the rest of my bottle of Military Special Vodka (yes, they DO make a type of liquor named "Military Special"... it sells for $4.35 per fifth at the Class Six) and mixed in four packets of "Electrolytes - 1 each - Cherry flavored" (I swear, the packet says that). It looked like an eerie Kool-Aid with sparkles in it (that was the electrolytes, I guess). Could've been leftover sand from Egypt. I lit two candles, put a vase of wildflowers in the middle, and set the table with my best set of Ralph Lauren Academy-series China (that stuff is EXPENSIVE... my set of eight place settings cost me over six hundred dollars on sale at the Camp Lejeune PX), and put the alcoholic drink in a crystal wine decanter. She came over, and I had some appetizers ready - made of MRE spaghetti-with-meatballs, set in small cups. She saw the dinner, saw the food, and said "This looks INCREDIBLE!!!"

We dug in, and she loved the food. Throughout the meal, she kept asking me how long it took me to make it, and kept remarking that I obviously knew a thing or two about cooking fine meals. She kind of balked at the make-shift "wine" I had set out, but after she tried it I guess she liked it because she drank four glasses during dinner.

At the end of the main course, when I served the dessert, she squealed with delight at the "chocolate mousse" I had made. Huh? Chocolate what? Okay... yeah... it's chocolate mousse. Took me HOURS to make... Yup!

Later on, as we were watching a movie, she excused herself to use my rest room. While she was in there I heard her say softly to herself "uh oh" and a resounding but petite fart punctuated her utterance of dismay. Let the games begin. She sprayed about half a can of air freshener (Air Freshener, 1 each, Orange scent. Yup. The military even makes smell-good) and returned to the couch, this time with an obviously pained look.

After ten more minutes she excused herself again, and retreated to the bathroom for the second time. I could hear her say, "What the hell is WRONG with me?" as she again sent flatulent shockwaves into the porcelain bowl. This time they sounded kinda wet, and I heard the toilet paper roll being employed, and again LOTS more air freshener.

Back to the couch. She smiles meekly as she decides to sit on the chair instead of next to me. She sits on my chair, knees pulled up to her chest, kind of rocking back and forth slightly. Suddenly, without a word, she ROCKETED up and FLEW to the bathroom, slammed the door, and didn't come out for thirty minutes.

I turned the movie up because I didn't want her to hear me laughing so hard that tears were streaming down my cheeks. She came out with a slightly gray pallor to her face, and said "I am SOOOOOO sorry. I have NO idea what is wrong with me. I am so embarrassed. I can't believe I keep running to your bathroom!"

I gave her an Imodium AD, and she finally settled down and relaxed. Later on she asked me again what I had made for dinner, because she had enjoyed it so much. I calmly took her into the kitchen and showed her all the used MRE bags and packets in the trash can. After explaining to her that she had eaten roughly nine thousand calories of "Marine Corps Field Rations" she turned stark white, looked at me incredulously, and said "I ate nine *thousand* calories of dehydrated food that was made three *years* ago?"

After I admitted it she grabbed her coat and keys and took off without a word. She called me yesterday. Seems she couldn't shit for five days, and when she finally did the smell was so bad her roommate could smell it from down the hall. She also told me she had been working out nonstop to combat the high caloric intake, and that she never wanted me to cook dinner for her again unless she was PERSONALLY present and supervising.

It was a fun date. She laughed about it eventually and said that it was the first time she'd ever crapped in a guy's house on a date. She'd been so upset by it she was in tears in the bathroom - while I had been in tears on the couch.

I know... I'm an asshole - but it was still a funny night.

# YOU CAN TELL A MARINE

**"Jurph"**

*If you have ever wondered how we are viewed by civilians, this may give you some idea.*

You can tell a Marine...but you sure can't tell him much. Like every joke, it has its roots in the truth. The first part does, anyhow. I don't care to test a Marine's sense of humor. United States Marines are amazingly easy to spot in a crowd. Other members of the U.S. military aren't quite as simple to identify, but you'd be surprised how easy it is once you start looking.

This node is intended to be a short guide to identifying people from a distance who may be U.S. military personnel, and is admittedly skewed towards identifying men since the women in the military tend to retain their fashion sense even after training.

First of all, look at the individual. Everyone in the military is required to keep their hair cut within regulations - with the possible exception of Special Forces... but if you're dealing with them, you'd better already know most of this stuff. The haircut will not necessarily be the ubiquitous white wall or jarhead haircut, but some sort of high and tight or buzz cut. A male in the military will have short hair, and will rarely have any facial hair. The exception is a small mustache, trimmed so that it doesn't extend past the corners of the mouth.

The clothing worn off-duty by military personnel tends to be conservative, too. I call the following outfit the "official off-duty uniform" because it's such a dead giveaway. Look for khakis (or "dressy" blue jeans) worn with a collared golf shirt. The shirt is almost certain to be tucked in. Casual shoes will be brown or black (athletic shoes are less common) and shined if appropriate. Navy and Marine Corps personnel may have a white crew-neck T-shirt showing underneath the open button on their collar. Air Force and Army personnel will not.

Because of the etiquette surrounding the salute, many military members will carry a briefcase in their left hand rather than their right. This frees the right hand for saluting and holding doors. Likewise, a backpack will be worn over the left shoulder. Look for a watch worn on the left wrist (to indicate right-handedness) before you jump to conclusions, though. Men in the military will not have pierced ears, and unless they're out clubbing in something revealing you probably won't see their tattoos.

We hang out with the people we work with, by and large, so look for military members to travel in packs just as much as any other social group. Four guys with the same haircut and the aforementioned outfits hanging out with a girl who's clearly at ease? Odds are good they're military (or gay). Listen for lots of acronyms in their speech. Like most Americans, military members tend to speak a little more loudly than absolutely necessary, and sometimes a little eavesdropping can tell you everything you need to know.

The U.S. military is one of the most integrated workplaces in the country - look for an astonishing mix of races among the men and women. If the men are out with their wives, look for spouses from overseas - especially Japan, Korea, Germany and Saudi Arabia (less common).

That's all I can offer, but I will throw this last tip out - attitude. There's a difference in their gait, in the way they deal with waitresses and clerks (when's the last time you called the Starbucks barista "ma'am"?), and in their situational awareness. Whether you're using this guide to avoid them or pick them up, I hope the advice I've offered helps you find the military man you're looking for.

By the way, if you happen to identify someone as being a Marine… try not to piss him off.

**This was originally posted in *Everything2*, an online collection of user-submitted writings about, well, pretty much everything.**

# CULT OF PERSONALITY

Chuck Goudie

We're a nation at war, and a golfer commandeered the country's airwaves to apologize for cheating on his wife. The only thing missing from Tiger Woods' public self-stoning was a scarlet "A" on his sport coat - but this isn't about the puffy, somber Woods who took over TV and radio last Friday. It's about a young man who couldn't get to a TV to watch. Josh Birchfield was his name. For thirteen minutes as the world stood still while Tiger talked, Josh was preoccupied. He was busy dying for his country.

The twenty-four-year-old United States Marine from Westville, Indiana was in Afghanistan, a long way from home and a world away from Tiger Woods. As Woods kept repeating how sorry he was for bedding countless women and violating his wedding vows, Lance Corporal Birchfield was working on the biggest military offensive since the 2001 U.S.-led invasion of Afghanistan.

Birchfield, a fresh-faced Marine who just a couple years prior was a high school baseball player and remained a devoted Chicago Cubs fan to the end, was killed in Marjah on a Friday morning. The Afghan city 360 miles southwest of Kabul had been a Taliban stronghold for several years, and the Marines were part of a sizable force which continued to dislodge Taliban fighters.

In the LaPorte County town of Westville, with a

281

population of only twenty-one hundred, word spread quickly once Birchfield's parents were notified of their son's death. Even on that Friday night, as analysts, advisers and experts dissected the statement read by Tiger Woods before he headed back for more sex counseling, the Birchfields hadn't been informed exactly how their son died.

Only two years into the Marine Corps, was his last moment on earth facing a rocket-propelled grenade, or had it been an improvised explosive device buried by an extremist on the side of a road? The uniformed delegation that came to inform them didn't know.

By Saturday, it seemed likely that Josh Birchfield was the Marine referred to by NATO officials when they reported one fatality on Friday during a small-arms skirmish in Marjah.

Down in Westville, topic number one wasn't the location of Tiger Woods' next golf tournament appearance. Whether Woods might play in the Masters didn't matter to the dozens of people who gathered in the Blackhawk Inn bar. They were too busy trying to plan a funeral.

"The sacrifice he made for his country is definitely hitting me pretty hard," Staff Sergeant Brandun Schweizer, one of Birchfield's good friends, told WSBT-TV. Schweizer, who served three tours in Iraq, described his dead friend as "a great, great individual. A truly great young man. He said he was tired of not going anywhere with his life. He wanted to get out there and do something meaningful."

"He had a fine life here. He was doing well," another friend, Steve Bachman, told WSBT. "He had a good job, good family, good people around him… but he saw people with families that had their children back here living with grandparents while they're over there fighting for us. He was just a regular, single guy, twenty-two-years-old, and he said, 'What am I doing? Why don't I go help?'"

You didn't hear Birchfield make that proclamation on television. For those who do our heavy lifting, there are no paparazzi. He just signed up, did the training and took the oath.

Some people will remember Friday, February 19, 2010 as the day Tiger Woods said, "It's not what you achieve in life that matters. It's what you overcome." I'll remember it as the day that Marine Lance Corporal Josh Birchfield did more than talk. He died a real American hero. No apology necessary.

**Chuck Goudie, whose column appears each Monday, is the chief investigative reporter at ABC 7 News in Chicago. The views in this column are his own and not those of WLS-TV.**

# YOU JOINED US

Jordan Blashek

**"Ultimately... (it) was a duty I could not, and should not, leave for others to assume."**

*The short essay below is by Jordan Blashek, Princeton 2009, who decided to turn down acceptance to medical school to join the Marine Corps and enter Officer Candidate School, from which he graduated in December of 2009.*

"You Joined Us" - that phrase is carved into a steel plaque which tauntingly guards the entrance to the Officers' barracks at Camp Barrett in Quantico, Virginia. As I hobbled inside, exhausted from another fifteen-hour day, my roommate half-jokingly pointed to the plaque, "Why did we do that, again?" I smiled. Today had been a long day. Waking at four AM, we spent the next nine hours outside in the pouring rain learning hand-to-hand combat and outdated bayonet techniques. Without warming layers, hats or gloves, our hands quickly went numb and our bodies started shaking uncontrollably in the thirty-degree temperature. Finally we were sent back inside to clean our rifles, which must be spotless before we can wash off our bodies. As eight PM rolled around and we were still cleaning on a Friday night - when my high school and college friends were out at Happy Hours - I thought about that plaque on the wall. Why exactly did I join, again?

# Been There, Done That... Got the T-Shirt!

It's a question I have tried to answer many times for my family and friends, but never feel as though I have fully conveyed my reasons. I made the decision to join the U.S. Marine Corps at the start of my senior year at Princeton, turning down an acceptance to medical school in the process. I kept the decision to myself until I broke the news to my shocked parents over Christmas Break. I ran through the litany of justifications for them. I wanted to serve my country. I wanted the camaraderie and the pride of being in the Marine Corps brotherhood. I needed the challenge to test my true capabilities and strength. I would receive the best leadership training on the planet, which would help me in any future career I chose. I wanted adventure and the chance to be a part of history in Iraq or Afghanistan. I wanted to exude that same confidence that I saw in every Marine officer I have met. Whether I convinced them or not, in the end none of these "reasons" alleviated my parents' understandable anxiety.

When I told my plans to anyone else, I felt as though I was talking to a brick wall - the military, especially the Marine Corps, was simply outside their reality. My closer friends would nod their heads and say something to the effect of "Wow, that's cool," but since I was the perennial flake of the group most did not take my decision very seriously - and to be honest, even I was not quite sure that I would follow through with the choice. In the comfort of my college dorm the decision to become a Marine Corps officer seemed glamorously abstract, however on October 1, 2009 my decision suddenly became very real when I arrived at the

Marine Corps' Officer Candidate School (OCS) in Quantico.

My OCS experience was surreal. Along with 407 other "candidates" - all college graduates with newly shaved heads - I ran around for ten weeks carrying an M16 rifle while the Marine Corps' famous drill instructors screamed increasingly creative insults at us. In reality, we were beginning the painful, yet deliberate process of transforming from civilians into Marine officers through some of the most intense training that exists in the U.S. military. Meanwhile, the drill instructors continually evaluated our leadership potential as part of the time-honored tradition whereby enlisted Marines select the officers who will eventually lead them in combat. After nearly half of the officer candidates were dropped or dropped out on their own, we emerged from OCS standing a little taller and a little straighter on graduation day - December 11, 2009. That afternoon I raised my right hand to swear the oath of office and receive my commission as a second lieutenant. That oath obligates me to serve a minimum of four years in uniform.

Ultimately, I joined the Marine Corps because I believe that officers bear the most solemn responsibility in our nation, and that was a duty I could not, and should not, leave for others to assume. To say that I wanted that responsibility is not quite right, because being a Marine officer is not about one's self, wants or needs. It is about guiding the young eighteen and nineteen-year-old Marines fighting this country's wars on our behalf. I decided that serving them was the highest honor and responsibility I could have at this point in my life. As one speaker at my commissioning ceremony explained, "As second lieutenants, you must have a strong sense of the great responsibility of your office. The resources which you will expend in war are human lives. This is not about *you* anymore. This is about the young Marines who will place their lives in your hands. It is your job to take care of them, even when that means placing them in mortal danger. That awesome responsibility - the weight which now rests on you - is reflected in those gold bars which you will soon place on your shoulders."

That is why the plaque hangs in every portal through which we pass – "You Joined Us." We chose to bear this responsibility and we must make absolutely sure we are prepared to fulfill it, because young American lives are at stake. If that means being cold and miserable, studying for ungodly hours and going for days without sleep, then so be it. That is the price of the salute we receive from our Marines.

Five months into my service commitment, I have not regretted my decision for a moment. I already have unforgettable memories from my experience and new friendships with diverse and exceptional peers from all over the country. We have had moments of pure fun together and laughed harder than I ever thought possible. We have also been humbled by the stories and portraits of brave Lieutenants - those who fought and died after roaming the very halls where we now stand and their portraits hang. Most of all, I am immensely proud to bear the title of United States Marine, an honor that I will carry with me my entire life. Semper Fi.

# A MOMENTARY LAPSE
## *Of Reason*

Lex McMahon

"Sir, no sir, this recruit does not think that Drill Instructor Staff Sergeant Carpenter is an asshole, Sir!" This phrase became my mantra for twelve long weeks in 1991. The Drill Instructor in *Full Metal Jacket* had nothing on the sadistic bastard who pounded me and crushed my ego - but I'm getting carried away.

Let's start from the beginning. When I joined the Corps I was two hundred and thirty pounds of mostly muscle. I'd just completed my first season of college football, and was the easiest sell the recruiter ever had. I walked into his office and said, "Sign me up, I want to be a grunt. I can leave tomorrow if need be. I'm ready to go. I'm a big dude. I can handle it."

My recruiter, Sergeant Mack, flashed a grin. During the month I was waiting to ship to boot camp, Mack filled my head with delusions of grandeur. He assured me that with my gung ho attitude I'd surely be the next Chesty Puller - or at least graduate as company honor man.

Emboldened by Mack's assurances that I was destined for my own statue next to the Iwo Jima Memorial, I stepped off the bus at Marine Corps Recruit Depot San Diego with an arrogant swagger and placed my feet on the yellow footprints awaiting what was certain to be immediate recognition of how badass I really was. Of course, the DI's did not quite see it my way. Over the course of the first sleep deprived days, where it seemed that I could do nothing right, I began to question Mack's motivation for praising my abilities.

By the time the receiving process ended and my platoon was finally formed, I no longer had that swagger and cockiness. The first few days had been rough, but I figured it couldn't get any worse, right? And Bill Clinton did not have sexual relations with that woman, Monica Lewinsky, either.

It was time to meet the training cadre and the DI who would quickly become my nemesis, Drill Instructor Staff Sergeant Carpenter, a Force Recon Marine who I was convinced just by looking at him could drink napalm and piss pure fire. He was the living incarnation of John Wayne.

On one of the first training days my platoon was taken to the PT field for an introduction to log drills, where a team of five or six recruits join together to conduct exercises using a five hundred pound log. In the civilian world it would make sense to put teams together of people of similar size so as to spread the workload evenly - but this was the United States Marine Corps, where adversity routinely dropped from the heavens. I was placed on a log with the newest members of what the DI's called the "Midget Militia." Not one of the recruits was over 5'4" tall or weighed more than 145 pounds. To add insult to injury, I was placed on the end of the log, which resulted in me having to do the lion's share of the work. As I grunted out each repetition Drill Instructor Carpenter began hurling insults at me.

"Recruit Fat Body (that's me), you're not working hard

enough, you're letting the Midget Militia down! Are you going to let your fellow Marines down in combat? I think so! You don't deserve to join my beloved Corps!"

That's when it happened. The words came out before I could stop them. Like the lead singer from the Dixie Chicks who bashed President Bush, I said something which would reverberate through my entire life - and I instantly regretted it. I looked Drill Instructor Carpenter square in his Charlie Manson gaze and said, "You're an asshole!"

Oh. Dear. Lord. The earth cracked open. People began falling into the deep chasm which threatened the very survival of the human race, and yet Carpenter and his minions felt the need to correct me instead of running for cover. In the midst of the hellish thrashing I endured several thoughts occurred to me:

- Did I really just say that?
- In a split second I've made my stay on the Depot infinitely harder.
- I wonder what my friends back home are doing right now? (I bet they are at Fat Burger).
- I miss Fat Burger. Especially when the cheese melts over the sides of the bun and you have to pry it off the plate with a sharp knife because there ain't no way in hell I'm not eating it.
- Isn't it elk season?
- Someone just broke wind.
- Snap out of it! It's time to suck it up and prove I deserve to be a Marine, despite my catastrophic loss of bearing.

It was in this moment of clarity that I resolved to never be broken again. I was determined to be the toughest son of a bitch in the valley!

At the conclusion of my epic thrashing, I thought, "Well that pretty much sucked, but it's over. I've paid my penance and atoned for my sins." Ah, to be young and naive again.

Drill Instructor Carpenter saw my insult as an opportunity to teach other recruits how not to be the shit bird that I was. At the start of every class throughout the remaining twelve weeks of boot camp, he would call me to attention in front of the rest of the platoon or company and ask if I thought he was still an asshole. My robotic response was always the same: "Sir, no sir, this recruit does not think that Drill Instructor Staff Sergeant Carpenter is an asshole, Sir!"

The payback did not stop with the public inquisitions. Every time I turned around there was Drill Instructor Carpenter, whispering in my ear that he was going to rip my eyes out and use my empty eye sockets as his personal spittoon. In the middle of the night, I would awake to find him hovering millimeters from my face ready to administer some "old Corps" remedial training.

Despite, or more likely because, of Drill Instructor Carpenter's constant pressure I graduated as honor man and was meritoriously promoted. On graduation day, basking in the euphoria of the moment with my family, he approached with a grin and asked, "So Marine, am I still an asshole"? I thought for a moment and responded, "Yes, Staff Sergeant. You are, but thank you. You made me the Marine I am today." We shook hands and he walked away. I would never see Carpenter again, but I have thought of him often. The lessons he imparted helped me persevere and thrive when the bullets started flying in Somalia. Discipline will not only separate you from the long haired hippies who saunter through life with blinders on and not a care in the world, but it will also save your life - even if it takes an asshole to make you realize it.

# "MARINESTAN"

**Victor Davis Hanson**

HBO's ten-part series on the Pacific campaign of World War II just ended. That story of island-hopping was mostly about how the old breed of U.S Marines fought diehard Japanese infantrymen face-to-face in places like Guadalcanal, Tarawa, Saipan, Peleliu, Iwo Jima, Guam and Okinawa.

We still argue whether it was smart to storm those entrenched Japanese positions or whether all those islands were strategically necessary, but even so no one can question the Marine Corps' record of having defeating the most savage infantrymen of the age, thereby shattering the myth of Japanese military invincibility.

Since WWII, the Marines have turned up almost anywhere that America finds itself in a jam against supposedly unconquerable enemies - in bloody places like Inchon and the Chosin Reservoir in Korea, at Hue and Khe Sanh during the Vietnam War, at the two bloody sieges of Fallujah in Iraq, and now in Afghanistan.

Over the last two centuries, two truths have emerged about the Marine Corps. One, they defeat the toughest of America's adversaries under the worst of conditions, and two, periodically their way of doing things - and their eccentric culture of self-regard - so bothers our military planners that some higher-ups try either to curb their

293

independence or end the Corps altogether.

After the Pacific fighting, Secretary of Defense Louis Johnson wanted to disband the Marines Corps. "What good were amphibious landings in the nuclear age?" Johnson asked. His boss, President Harry Truman, agreed. He didn't like the cocky Marines either.

Then came Korea - and suddenly the Pentagon wanted *more* Marines. The fighting against hard-core North Korean and Communist Chinese veterans was as nasty as anything seen in three millennia of organized warfare, and the antiquated idea of landing on beaches proved once again a smart way of outflanking the enemy.

The Marines survived the Korean and Chinese armies - as well as Louis Johnson and Harry Truman - and continued to carve out their own logistics, air-support and tactical doctrine. Marine self-sufficiency was due to lingering distrust of the other services dating back to the lack of air and naval support in World War II, and to Marine paranoia that the other services liked their combative spirit - but not their independence.

We are once again seeing one of those periodic re-examinations of the Corps, and this time the old stereotype of the lone-ranger, gung-ho Marines supposedly doesn't fit too well with fighting sophisticated urban counterinsurgency under an integrated, international command. After all, America is fighting wars in which we rarely hear of the number of enemy dead, but a great deal about the need to rebuild cities and infrastructure. In Afghanistan, there have even been rumors about a new medal for "courageous restraint" that would honor soldiers who hesitated pulling the trigger against the enemy out of concern about harming civilians. Such an award might work for the other services, but I don't think too many Marines will be getting that one.

The Marines are now starting to redeploy to Afghanistan from Iraq and are building a huge base in Delaram. They plan to win over southern Afghanistan's remote, wild Nimruz province that heretofore has been mostly a no-go Taliban stronghold. While NATO forces concentrate on Afghanistan's major cities, the Marines think they can win over local populations their way, take on and defeat the Taliban, and bring all of Nimruz back from the brink - with their trademark warning of "no better friend, no worse enemy."

So once again, the Marines are convinced that their own ingenuity and audacity can succeed where others have failed. And once again, not everyone agrees. The U.S. Ambassador to Afghanistan, retired three-star Army General Karl W. Eikenberry, reportedly made a comment about there being forty-one nations serving in Afghanistan - and a *forty-second* composed of the Marine Corps. One unnamed Obama administration official was quoted by the *Washington Post* as saying, "We have better operational coherence with virtually all of our NATO allies than we have with the U.S. Marine Corps."

Some officials call the new Marine enclave in Nimruz Province "Marinestan" - as if it is something out of a Kipling or Conrad novel, in which the Marines have gone rogue to set up their own independent province of operations. Yet once again, it would be wise not to tamper with the independence of the Marine Corps, given that its methods of training, deployment, fighting, counterinsurgency and conventional warfare usually pay off in the end.

The technological and political face of war is always changing, but its essence - organized violence to achieve political ends - is no different from antiquity. Conflict will remain the same as long as human nature does as well. The Marines have always best understood that, and from the Corps' initial mission against the Barbary Pirates to the recent battles in Fallujah, Americans have wanted a maverick Marine Corps - a sort of insurance policy that kept them safe... just in case.

**Victor Davis Hanson is a military historian, columnist, political essayist and former professor whose writings can be found at www.victorhanson.com**

# SEEING THE FALLEN HOME

Colleen M. Getz

His name was Marine Lance Corporal Justin Wilson - although I did not know it when his life brushed mine on March 25 at Ronald Reagan Washington National Airport. Lance Corporal Wilson was not there in the terminal that afternoon - at age twenty-four and newly married, he had been killed in Afghanistan on March 22 by a roadside bomb. A coincidence of overbooked flights led our lives to intersect for perhaps an hour, one I will never forget.

I did not meet his family that day at the airport either, although we were there together that evening at the gate, among the crowd hoping to board the oversold flight. I did not know that I had a boarding pass and they did not. I did not know they were trying to get home to hold his funeral, having journeyed to Dover, Delaware to meet his casket upon its arrival from Afghanistan.

I also did not know they already had been stuck for most of the day in another airport because of other oversold flights - but I did not need to know this to realize what they were going through as the event unfolded and to understand the larger cause for it. No matter how we as a nation have relearned the lesson forgotten during Vietnam - that our military men and women and their families deserve all the support we can give them - despite our nation's fighting two wars in this decade, it is all too easy for most of us to live

297

our lives without having the very great human cost of those wars ever intrude.

But it did intrude heartbreakingly that day at the airport gate. It began simply enough, with the usual call for volunteers - anyone willing to take a later flight would receive a five hundred dollar flight voucher. Then came the announcement none of us was prepared to hear. There was, the airline representative said, a family on their way home from meeting their son's body as it returned from Afghanistan, and they needed seats on the flight - and there they were, standing beside her as she looked at us, waiting to see what we would decide. It wasn't a hard decision for me, since my plans were easily adjusted. I volunteered, as did two women whom I later learned sacrificed important personal plans.

But we three were not enough. Six were needed, so we stood there watching the family - dignified and mute, weighed with grief and fatigue - as the airline representative repeatedly called for assistance for this dead Marine's family. No one else stepped forward. The calls for volunteers may have lasted only twenty or thirty minutes, but it seemed hours. It was almost unbearable to watch, yet to look away was to see the more than a hundred other witnesses to this tragedy who were not moved to help. Then it did become unbearable when, in a voice laced with desperation and tears, the airline representative pleaded, "This young man gave his *life* for our country. Can't any of you give your *seats* so his family can get home?" Those words hung in the air, and finally enough volunteers stepped forward.

I had trouble sleeping that night. I could not get out of my mind the image of the family or the voice pleading for them. When I met my fellow volunteers the next morning at the airport, I found I was not alone. One had gone home and cried, and another had awakened at three AM. All of us were angry and ashamed that our fellow passengers had not rushed to aid this family and consequently had forced them to be on public display in their grief. We worried that this indifference to their son's sacrifice added to their sorrow.

It turned out my destination was his hometown, so I was able to learn his name and more. I learned he had been a talented graffiti artist and had married his sweetheart, Hannah, the day before he deployed to Afghanistan. They planned a big wedding with family and friends for after he returned home. I learned how proud he was to become a Marine in January 2009. I learned that he and his fellow Marines liked to give the candy they received from home to Afghan children. In sum, I learned that he was the kind of honorable, patriotic young person we want defending our country and how great our loss is that he had to give his life in doing so.

I posted a message to his family on the online condolence book. I told them I was sorry for what they went through in trying to see their son's body home, but said because of it many more people were going to have heard of Justin and his dedication to his country because I was going to tell everyone I knew about what I had witnessed and tell them his name - and I have.

I thought that was enough, until I received a thank-you

note from Lance Corporal Wilson's father-in-law. It was a completely humbling experience. He wrote that he was glad I had been able to learn about Justin, and he wanted me to know that Justin "served knowing the risks, but felt it was his obligation and privilege to serve his country." At that moment, I realized in this day of an all-volunteer military and a distant war which touches so few of our lives directly, more people should hear the story of Lance Corporal Wilson and his family.

I've thought a lot about what happened that day in the airport, and I choose to believe my fellow passengers were not unfeeling in the face of a Marine's death and a family's tragedy. They were just caught off guard - they were totally unprepared to confront the fierce consequences of the war in Afghanistan on their way to Palm Beach on a sunny afternoon - and I believe it was for this reason people did not rush to the podium to volunteer their seats. It was not that they did not want to, and it was not that they did not think it was the right thing to do. Rather, it was because they were busy trying to assimilate this unexpected confrontation with the irrevocable cost of war and to figure out how to fit doing the right thing into their plans - to fit it into their lives not previously touched by this war. In the end, enough of us figured out how to do the right thing, and it turned out as well as such a painful situation could.

But still I wonder. Barring some momentous personal event that necessitated a seat on that flight, how could any of us even have hesitated? How could we have stopped to weigh any inconvenience to our plans against the sacrifice

Justin Wilson and his family had made for our country? In such circumstances, it is not a question of recognizing the *right* thing to do - we should know it is the *only* thing to do.

From what I have learned of him, in his short life Lance Corporal Justin Wilson created a legacy of courage and patriotism that will not be forgotten by those who knew him. I hope there's a greater legacy as well. I hope through this account of his family's struggle to see him home, if ever again the war intrudes unbidden on my life or yours, we will know what we must do - and in their honor, and for all those who serve and sacrifice - we will do it.

**Colleen M. Getz works in the NATO policy office of the Department of Defense. This article originally appeared in the *Washington Times* on May 31, 2010.**

# THE LOST STANZA

Some of you oldtimers... and I'm talking *serious* Old Corps here... may know that up until approximately 1930 there were four stanzas to *The Marines' Hymn*. The original third stanza was removed so as to not hurt the feelings of the German people, although I find it hard to understand why anyone would be concerned with offending a nation which has plunged the entire world into war - twice.

Here is the original third stanza, as it was printed in the 4th Marine Regiment's *Legation Guard News* in Peking, China (circa 1930).

> *When we were called across the sea,*
> *To stand for home and right.*
> *With the spirit of the brave and free,*
> *We fought with all our might.*
> *When we helped to stem the German drive,*
> *They say we fought like fiends,*
> *And the French rechristened Belleau Wood,*
> *For the United States Marines.*

I like it, especially the part about "fighting like fiends" and "rechristening Belleau Wood" - and think we should start a petition to bring it back. What do *you* think?

# THE LAST SIX SECONDS

*On November 13, 2010 Lieutenant General John Kelly, USMC gave a speech to the Semper Fi Society of St. Louis, just four days after his son, Lieutenant Robert Kelly, USMC was killed by an IED while on his third combat tour. During his speech General Kelly spoke about the dedication and valor of the young Marines who step forward each and every day to protect us, and never mentioned the loss of his own son. He closed the speech with the moving account of the last six seconds in the lives of two young Marines who died with their rifles blazing to protect their brother Marines.*

Nine years ago four commercial aircraft took off from Boston, Newark, and Washington. Took off fully loaded with men, women and children - all innocent, and all soon to die. These aircraft were targeted at the World Trade Towers in New York, the Pentagon, and likely the Capitol in Washington, D.C. Three found their mark. No American alive old enough to remember will ever forget exactly where they were, exactly what they were doing, and exactly who they were with at the moment they watched the aircraft dive into the World Trade Towers on what was, until then, a beautiful morning in New York City. Within the hour three thousand blameless human beings would be vaporized, incinerated or crushed in the most agonizing ways imaginable. The most wretched among them - over two hundred - driven mad by heat, hopelessness and utter desperation, leapt to their deaths from a thousand feet above Lower Manhattan. We soon learned hundreds more were murdered at the Pentagon, and in a Pennsylvania field.

Once the buildings had collapsed and the immensity of the attack began to register most of us had no idea of what to do, or where to turn. As a nation, we were scared like we had not been scared for generations. Parents hugged their children to gain as much as to give comfort. Strangers embraced in the streets stunned and crying on one another's shoulders seeking solace, as much as to give it. Instantaneously, American patriotism soared not "as the last refuge" as our national-cynical class would say, but in the darkest times Americans seek refuge in family, and in country, remembering that strong men and women have always stepped forward to protect the nation when the need was dire - and it was so God awful dire that day - and remains so today.

There was, however, a small segment of America that made very different choices that day... actions the rest of America stood in awe of on 9/11 and every day since. The first were our firefighters and police, their ranks decimated that day as they ran towards - not away from - danger and certain death. They were doing what they'd sworn to do - "protect and serve" - and went to their graves having fulfilled their sacred oath. Then there were the Armed Forces, and I know I am a little biased in my opinion here, but the best of them are Marines. Most wearing the Eagle, Globe and Anchor today joined the unbroken ranks of

American heroes after that fateful day not for money, or promises of bonuses or travel to exotic liberty ports, but for one reason and one reason alone - because of the terrible assault on our way of life by men they knew must be killed and extremist ideology that must be destroyed. A plastic flag in their car window was not their response to the murderous assault on our country. No, their response was a commitment to protect the nation swearing an oath to their God to do so, to their deaths. When future generations ask why America is still free and the heyday of Al Qaeda and their terrorist allies was counted in days rather than in centuries as the extremists themselves predicted, our hometown heroes - soldiers, sailors, airmen, Coast Guardsmen, and Marines - can say, "because of me and people like me who risked all to protect millions who will never know my name."

As we sit here right now, we should not lose sight of the fact that America is at risk in a way it has never been before. Our enemy fights for an ideology based on an irrational hatred of who we are. Make no mistake about that, no matter what certain elements of the "chattering class" relentlessly churn out. We did not start this fight, and it will not end until the extremists understand that we as a people will never lose our faith or our courage. If they persist, these terrorists and extremists and the nations that provide them sanctuary, they must know they will continue to be tracked down and captured or killed. America's civilian and military protectors both here at home and overseas have for nearly nine years fought this enemy to a standstill and have never for a second "wondered why." They know, and are not afraid. Their struggle is your struggle. They hold in disdain those who claim to support them but not the cause that takes their innocence, their limbs, and even their lives. As a democracy - "We the People" - and that by definition is every one of us - we sent them away from home and hearth to fight our enemies. We are all responsible. I know it doesn't apply to those of us here tonight, but if anyone thinks you can somehow thank them for their service, and not support the cause for which they fight - America's survival - then they are lying to themselves and rationalizing away something in their lives... but more importantly, they are slighting our warriors and mocking their commitment to the nation.

Since this generation's "day of infamy" the American military has handed our ruthless enemy defeat-after-defeat but it will go on for years, if not decades, before this curse has been eradicated. We have done this by unceasing pursuit day and night into whatever miserable lair Al Qaeda, the Taliban and their allies, might slither into to lay in wait for future opportunities to strike a blow at freedom. America's warriors have never lost faith in their mission, or doubted the correctness of their cause. They face dangers everyday that their countrymen safe and comfortable this night cannot imagine - but this has always been the case in all the wars our military have been sent to fight. Not to build empires, or enslave peoples, but to free those held in the grip of tyrants while at the same time protecting our nation, its citizens and our shared values. And ladies and gentlemen, think about this - the only territory we as a people have ever asked for from any nation we have fought alongside, or against, since our founding, the entire extent of our overseas empire, is a few hundred acres of land for the twenty-four American cemeteries scattered around the globe. It is in these cemeteries where 220,000 of our sons and daughters rest in glory for eternity, or are memorialized forever because their earthly remains are lost forever in the deepest depths of the oceans, or never recovered from far flung and nameless battlefields. As a people, we can be proud because billions across the planet today live free, and billions yet unborn will also enjoy the same freedom and a chance at prosperity because America sent its sons and daughters out to fight and die for them, as much as for us.

Yes, we are at war, and are winning, but you wouldn't know it because successes go unreported, and only when something does go sufficiently wrong, or is sufficiently controversial, is it highlighted by the media elite that then sets up the "know it all" chattering class to offer their endless criticism. These self-proclaimed experts always seem to know better - but have never themselves been in the arena. We are at war and like it or not, that is a fact. It is not Bush's war, and it is not Obama's war, it is *our* war and we can't run away from it. Even if we wanted to surrender, there is no one to surrender to. Our enemy is savage, offers absolutely no quarter, and has a single focus and that is either kill every one of us here at home, or enslave us with a sick form of extremism that serves no God or purpose that decent men and women could ever grasp. Saint Louis is as much at risk as is New York and Washington, D.C., and given the opportunity to do another 9/11 our merciless enemy would do it today, tomorrow and every day thereafter. If, and most in the know predict that it is only a matter of time, he acquires nuclear, chemical or biological weapons, these extremists will use these weapons of mass murder against us without a moment's hesitation. These butchers we fight killed more than three thousand innocents on 9/11. As horrible as that death toll was, consider for a moment that the monsters that organized those strikes against New York and Washington, D.C. killed only three thousand not because that was enough to make their sick and demented point, but because he couldn't figure out how to kill thirty thousand, or three hundred thousand, or thirty million of us that terrible day. I don't know why they hate us, and I don't care. We have a saying in the Marine Corps, and that is "no better friend, no worse enemy, than a U.S. Marine." We always hope for the first - friendship - but are certainly more than ready for the second. If its death they want, its death they will get, and the Marines will continue showing them the way to hell if that's what will make them happy.

Because our America hasn't been successfully attacked again since 9/11 many forget because we want to forget... to move on. As Americans we all dream and hope for peace, but we must be realistic and acknowledge that hope is never an option or course of action when the stakes are so high. Others are less realistic or less committed, or are working their own agendas, and look for ways to blame past presidents or in some other way rationalize a way out of this war. The problem is our enemy is not willing to let us go. Regardless of how much we wish this nightmare would go away, our enemy will stay forever on the offensive until he hurts us so badly we surrender, or we kill him first. To him this is not about our friendship with Israel, or about territory, resources, jobs or economic opportunity in the Middle East. No, it is about us as a people. About our freedom to worship any God we please in any way we want. It is about the worth of every man, and the worth of every woman, and their equality in the eyes of God and the law. Of how we live our lives with our families, inside the privacy of our own homes. It's about the God-given rights of life, liberty, and the pursuit of happiness and "that all men are created equal, that they are endowed by their Creator with certain inalienable right." As Americans we hold these truths to be self-evident. He doesn't. We love what we have. He despises who we are. Our positions can never be reconciled. He cannot be deterred... only defeated. Compromise is out of the question.

It is a fact that our country today is in a life and death struggle against an evil enemy, but America as a whole is certainly not at war. Not as a country. Not as a people. Today, only a tiny fraction - less than one percent - shoulder the burden of fear and sacrifice, and they shoulder it for the rest of us. Their sons and daughters who serve are men and women of character who continue to believe in this country enough to put life and limb on the line without qualification, and without thought of personal gain, and they serve so that the sons and daughters of the other ninety-nine percent don't have to. No big deal though, as Marines have always been "the first to fight," paying in full the bill that comes with being free... for everyone else.

309

# Been There, Done That... Got the T-Shirt!

The comforting news for every American is that our men and women in uniform, and every Marine, are as good today as any in our history. As good as what their heroic, under-appreciated and largely abandoned fathers and uncles were in Vietnam, and their grandfathers were in Korea and World War II. They have the same steel in their backs and have made their own mark etching forever places like Ramadi, Fallujah, and Baghdad in Iraq, and Helmand and Sagin in Afghanistan, that are now part of the legend and stand just as proudly alongside Belleau Wood, Iwo Jima, Inchon, Hue City and Khe Sahn. None of them have ever asked what their country could do for them, but always and with their lives asked what they could do for America. While some might think we have produced yet another generation of materialistic, consumerist and self-absorbed young people, those who serve today have broken the mold and stepped out as real men, and real women, who are already making their own way in life while protecting ours. They know the real strength of a platoon, a battalion or a country that is not worshiping at the altar of diversity, but in a melting pot that stitches and strengthens by a sense of shared history, values, customs, hopes and dreams - all of which unifies a people making them stronger, as opposed to an unruly gaggle of "hyphenated" or "multi-cultural individuals."

And what are they like in combat in this war? Like Marines have been throughout our history. In my three tours in combat as an infantry officer and commanding general I never saw one of them hesitate, or do anything other than lean into the fire and with no apparent fear of death or injury

310

take the fight to our enemies. As anyone who has ever experienced combat knows when it starts, when the explosions and tracers are everywhere and the calls for the Corpsman are screamed from the throats of men who know they are dying - when seconds seem like hours and it all becomes slow motion and fast forward at the same time - and the only rational act is to stop, get down, save yourself - they don't. When no one would call them coward for cowering behind a wall or in a hole, a slave to the most basic of all human instincts - survival - none of them do. It doesn't matter if it's an IED, a suicide bomber, mortar attack, sniper, fighting in the upstairs room of a house, or all of it at once. They talk, swagger, and most importantly fight today in the same way America's Marines have since Tun Tavern. They also know whose shoulders they stand on, and they will never shame any Marine living or dead.

We can also take comfort in the fact that these young Americans are not born killers, but are good and decent young men and women who for going on ten years have performed remarkable acts of bravery and selflessness to a cause they have decided is bigger and more important than themselves. Only a few months ago they were delivering your paper, stocking shelves in the local grocery store, worshiping in church on Sunday, or playing hockey on local ice. Like my own two sons who are Marines and have fought in Iraq, and today in Afghanistan, they are also the same kids that drove their cars too fast for your liking, and played the God-awful music of their generation too loud, but have no doubt they are the finest of their generation. Like those who went before them in uniform, we owe them everything. We owe them our safety. We owe them our prosperity. We owe them our freedom. We owe them our lives. Any one of them could have done something more self-serving with their lives as the vast majority of their age group elected to do after high school and college but no, they chose to serve knowing full well a brutal war was in their future. They did not avoid the basic and cherished responsibility of a citizen - the defense of country - they welcomed it. They are the very best this country produces, and have put every one of us ahead of themselves. All are heroes for simply stepping forward, and we as a people owe a debt we can never fully pay. Their legacy will be of selfless valor, the country we live in, the way we live our lives, and the freedoms the rest of their countrymen take for granted.

Over five thousand have died thus far in this war, eight thousand if you include the innocents murdered on 9/11. They are overwhelmingly working class kids, the children of cops and firefighters, city and factory workers, school teachers and small business owners. With some exceptions they are from families short on stock portfolios and futures, but long on love of country and service to the nation. Just yesterday too many were lost, and a knock on the door late last night brought their families to their knees in a grief that will never, ever go away. Thousands more have suffered wounds since it all started, but like anyone who loses life or limb while serving others - including our firefighters and law enforcement personnel who on 9/11 were the first casualties of this war - they are not victims as they knew what they were about, and were doing what they wanted to do. The chattering class and all those who doubt America's intentions, and resolve, endeavor to make them and their families out to be victims, but they are wrong. We who have served and are serving refuse their sympathy. Those of us who have lived in the dirt, sweat and struggle of the arena are not victims and will have none of that. Those with less of a sense of service to the nation never understand it when men and women of character step forward to look danger and adversity straight in the eye, refusing to blink or give ground, even to their own deaths. The protected can't begin to understand the price paid so they and their families can sleep safe and free at night. No, they are not victims, but are warriors - your warriors - and warriors are never victims regardless of how and where they fall. Death, or fear of death, has no power over them. Their paths are paved by sacrifice, sacrifices they gladly make... for you. They prove themselves everyday on the field of battle... for you. They fight in every corner of the globe... for you. They live to fight... for you, and they never rest because there is always another battle to be won in the defense of America.

I will leave you with a story about the kind of people they are... about the quality of the steel in their backs... about the kind of dedication they bring to our country while they serve in uniform and forever after as veterans. Two years ago when I was the Commander of all U.S. and Iraqi forces, in fact on the 22nd of April 2008, two Marine infantry battalions - 1/9 "The Walking Dead" and 2/8 - were switching out in Ramadi. One battalion was in the closing days of their deployment going home very soon, the other just starting its seven-month combat tour. Two Marines, Corporal Jonathan Yale and Lance Corporal Jordan Haerter, twenty-two and twenty years old respectively, one from each battalion, were assuming the watch together at the entrance gate of an outpost that contained a makeshift barracks housing fifty Marines. The same broken down ramshackle building was also home to a hundred Iraqi police, also my men and our allies in the fight against the terrorists in Ramadi, a city until recently the most dangerous city on earth and owned by Al Qaeda. Yale was a dirt poor mixed-race kid from Virginia with a wife and daughter, and a mother and sister who lived with him and he supported as well. He did this on a yearly salary of less than twenty-three thousand dollars. Haerter, on the other hand, was a middle class white kid from Long Island. They were from two completely different worlds. Had they not joined the Marines they would never have met each other, or understood that multiple America's exist simultaneously depending on one's race, education level, economic status and where you might have been born. But they were Marines, combat Marines, forged in the same crucible of Marine training, and because of this bond they were brothers as close, or closer, than if they were born of the same woman.

The mission orders they received from the sergeant squad leader I am sure went something like, "Okay you two clowns, stand this post and let no unauthorized personnel or vehicles pass. You clear?" I am also sure Yale and Haerter then rolled their eyes and said in unison something like, "Yes

314

Sergeant," with just enough attitude that made the point without saying the words, "No kidding sweetheart, we know what we're doing." They then relieved two other Marines on watch and took up their post at the entry control point of Joint Security Station Nasser, in the Sophia section of Ramadi, al Anbar, Iraq.

A few minutes later a large blue truck turned down the alleyway - perhaps sixty or seventy yards in length - and sped its way through the serpentine of concrete jersey walls. The truck stopped just short of where the two were posted and detonated, killing them both catastrophically.

Twenty-four brick masonry houses were damaged or destroyed. A mosque a hundred yards away collapsed. The truck's engine came to rest two hundred yards away, knocking most of a house down before it stopped. Our explosive experts reckoned the blast was made of two thousand pounds of explosives. Two died, and because these two young infantrymen didn't have it in their DNA to run from danger, they saved a hundred and fifty of their Iraqi and American brothers-in-arms.

When I read the situation report about the incident a few hours after it happened I called the regimental commander for details, as something about this struck me as different. Marines dying or being seriously wounded is commonplace in combat. We expect Marines, regardless of rank or MOS, to stand their ground and do their duty and even die in the process, if that is what the mission takes - but this just seemed different. The regimental commander had just returned from the site and he agreed, but reported that there were no American witnesses to the event - just Iraqi police. I figured if there was any chance of finding out what actually happened, and then to decorate the two Marines to acknowledge their bravery, I'd have to do it as a combat award that requires two eyewitnesses and we figured the bureaucrats back in Washington would never buy Iraqi statements. If it had any chance at all, it had to come under the signature of a general officer.

I traveled to Ramadi the next day and spoke individually to a half-dozen Iraqi police, all of whom told the same story. The blue truck turned down into the alley and immediately sped up as it made its way through the serpentine. They all said, "We knew immediately what was going on as soon as the two Marines began firing." The Iraqi police then related that some of them also fired, and then to a man ran for safety just prior to the explosion. All survived. Many were injured... some seriously. One of the Iraqis elaborated and with tears welling up said, "If they'd run like any normal man would to save his life." Any normal man. "What he didn't know until then," he said, "and what he learned that very instant, was that Marines are not normal." Choking past the emotion he said, "Sir, in the name of God, no sane man would have stood there and done what they did." No sane man. "They saved us all."

What we didn't know at the time, and only learned a couple of days later after I wrote a summary and submitted both Yale and Haerter for posthumous Navy Crosses, was one of our security cameras, damaged initially in the blast, recorded some of the suicide attack. It happened exactly as the Iraqis had described it. It took exactly six seconds from when the truck entered the alley until it detonated.

You can watch the last six seconds of their young lives. Putting myself in their heads, I supposed it took about a second for the two Marines to separately come to the same conclusion about what was going on once the truck came into their view at the far end of the alley. Exactly no time to talk it over, or call the sergeant to ask what they should do. Only enough time to take half an instant and think about what the sergeant told them to do only a few minutes before... "let no unauthorized personnel or vehicles pass." The two Marines had about five seconds left to live.

It took maybe another two seconds for them to present their weapons, take aim and open up. By this time the truck was half-way through the barriers, and gaining speed the whole time. Here the recording shows a number of Iraqi police, some of whom had fired their AKs, now scattering like the normal and rational men they were - some running right past the Marines. They had three seconds left to live. For about two seconds more, the recording shows the Marines' weapons firing non-stop... the truck's windshield exploding into shards of glass as their rounds take it apart and tore in to the body of the son-of-a-bitch who is trying to get past them to kill their brothers - American and Iraqi - bedded down in the barracks totally unaware of the fact that their lives at that moment depended entirely on two Marines standing their ground. If they had been aware, they would have known they were safe... because two Marines stood between them and a crazed suicide bomber. The recording

317

shows the truck careening to a stop immediately in front of the two Marines. In all of the instantaneous violence Yale and Haerter never hesitated. By all reports and by the recording, they never stepped back. They never even started to step aside. They never even shifted their weight. With their feet spread should width apart they leaned into the danger, firing as fast as they could work their weapons. They had only one second left to live.

The truck explodes. The camera goes blank. Two young men go to their God. Six seconds. Not enough time to think about their families, their country, their flag or about their lives or their deaths, but more than enough time for two very brave young men to do their duty... into eternity. That is the kind of people who are on watch all over the world tonight - for you.

We Marines believe that God gave America the greatest gift he could bestow to man while he lived on this earth - freedom. We also believe he gave us another gift nearly as precious - our soldiers, sailors, airmen, Coast Guardsmen, and Marines - to safeguard that gift and guarantee no force on this earth can ever steal it away. It has been my distinct honor to have been with you here today. Rest assured our America, this experiment in democracy started over two centuries ago, will forever remain the "land of the free and home of the brave" so long as we never run out of tough young Americans who are willing to look beyond their own self-interest and comfortable lives and go into the darkest and most dangerous places on earth to hunt down, and kill, those who would do us harm.

# UNIFORM CODE OF *MARINE* JUSTICE

When I was a young snuffie we used to say "Join the Navy and see the world, join the Marine Corps and *police* it," because it seemed as if we were always policing up someone's brass, cigarette butts or something else - and sometimes that "something else" was *human* refuse.

Take for example what happened in November of 2010, when a Marine reservist collecting toys for kids in Augusta, Georgia was stabbed as he helped to nab a shoplifter.

Best Buy sales manager Orvin Smith told *The Augusta Chronicle* that man was seen on surveillance cameras putting a laptop under his jacket, and when confronted he became irate, knocked down an employee, pulled a knife and ran toward the door.

Unfortunately for the perpetrator, just outside were four Marines collecting toys for the "Toys For Tots" program. Smith said the Marines stopped the man, but he stabbed one of them, Corporal Phillip Duggan, in the back - but fortunately the cut was not severe.

The suspect was later transported to the local hospital with two broken arms, a broken leg, possible broken ribs, multiple contusions and assorted lacerations including a broken nose and jaw... injuries he sustained when he "fell" while trying to run away after stabbing the Marine. What a shame!

# ATHEISTS IN FOXHOLES

Lieutenant Clebe McClary is a medically retired Marine who is now one of the most recognized motivational and inspirational speakers in America. He proudly proclaims himself to be "in the service of the Lord's Army" and asserts that, to him, "USMC" means "U. S. Marine for Christ." There is much to be admired about this Marine... but at the same time he is representative of a growing problem in both the Marine Corps and the military in general.

In1968, during his nineteenth reconnaissance mission in Vietnam, McClary was critically wounded during an enemy attack and suffered the loss of an eye and an arm - and was told he would never walk again. Despite all that, he never lost the determination, dedication and courage to overcome his circumstances... and as a result of his bravery under fire and the concern he showed for his men was awarded Silver and Bronze Stars. The details of his last mission, and subsequent journey, are spelled out in his book *Living Proof.*

I have a great deal of admiration for Lieutenant McClary, just as I do for all Marines who have served bravely and honorably - and especially so when they somehow overcome a set of circumstances which would be life shattering to a lesser man. I also fully support whatever mechanism it is they use to get through those difficult challenges - whether it be family, religion, or even therapeutic needlepoint. That said, I find it greatly disappointing when a man who has served so valiantly, and has sacrificed so much in the name of freedom, chooses to denigrate the equally loyal service of others who do not share his religious views.

The subject of religion and the military has become a touchy one in this age of Islamic jihad, but the concept and practice of religious freedom in the United States Armed Forces dates back to the earliest days of this nation. The Constitution outlines the basic concept of religious freedom in the Bill of Rights, more specifically the First Amendment which specifies that "Congress shall make no law respecting an establishment of religion. Or prohibiting the free exercise thereof..."

All branches of the United States military are afforded the same rights to religious freedom as are American civilians, however members of the Armed Forces willingly surrender certain rights when it impinges on military discipline or the successful completion of an objective. This guarantee of religious freedom is codified for the Armed Forces in "Accommodation of Religious Practices Within the Military Services," which describes the commander's responsibility to provide for religious accommodation - for *everyone*.

The free exercise of religious freedom in the military has, by and large, followed the same path as American society in general - that is, as the understanding of free exercise expanded outside the military, so did it expand within the services. The growing embrace of religious pluralism can perhaps best be seen in the expansion of the Chaplaincy, whose role it is to provide for the free expression of religious belief by the troops. For example, not until the war with

Mexico in 1846 were Roman Catholics incorporated into the Chaplain Corps. Up until that time only Protestants served as chaplains, which put the United States at a propaganda disadvantage when fighting Catholic Mexico. In 1862 the word "Christian" was stricken from regulations governing the appointment of chaplains in order to allow for the appointment of Jewish chaplains, and that was brought about as a result of a request made to President Abraham Lincoln by the Board of Delegates of American Israelites.

Then during World War II Greek Orthodox chaplains were authorized to minister to members of the Eastern Orthodox Church, and in 1987 the Department of Defense registered the Buddhist Churches of America as an ecclesiastical endorsing agency and thus opened the door for Buddhist chaplains. In 1993 the first Muslim chaplain was added - yet another sign of America's growing religious diversity, and in recognition of the Armed Forces' Constitutional responsibility to meet the free expression needs of those in its ranks who hold minority religious views.

Religious freedom takes on an additional importance in the current international environment, where religious motivations are an increasing rationale for waging conflict. At a time when the United States is encouraging greater religious freedom in Muslim nations, it is imperative for us to show by example that religious pluralism is a viable and preferred option. Any sign of hypocrisy in Unites States policy, official or otherwise, toward the free exercise of religion - including freedom *from* religion - within the military makes it more difficult to convince others to follow our lead.

Unfortunately, evangelical extremists such as Lieutenant McClary do not support the practice of other religions - which is, as I previously demonstrated, one of the principles this Nation was founded upon. In fact, he is even critical of other Christians who do not practice *his particular brand* of faith... and when it comes to those who do not adhere to any religion at all, the rhetoric becomes downright nasty.

The fact of the matter is non-theistic service members serve honorably around the world, and always will. For example Pat Tillman - the football player turned Army Ranger who gave up a lucrative career in the wake of 9-11 and ended up making the ultimate sacrifice for our country - was an atheist. In the eyes of people like Clebe McClary he is a heathen who deserves to burn in hell. In my eyes, he was a Patriot. How do you see him?

Now, with growing advocacy for gays and women in the military, this group has become the last unprotected minority. The non-theistic - whether they be atheist, humanist, agnostic, freethinker, or other secular minority - have served with just as much valor as anyone, but even so are discriminated against for not being "believers." To be honest, I have always had a difficult time reconciling the phrase "praise the Lord and pass the ammunition" with the Sixth Commandment... you know, the one which says "Thou Shalt Not Kill." The obvious rationalizations are to say it okay to kill while doing the Lord's work, or to reinterpret the Biblical translation to read "murder" rather than "kill," but in doing so Christians are opening a whole new can of worms. Remember, the justification Islamo-fascist Jihadist suicide bombers cite while slaying us infidels is they are doing Allah's work, or carrying out the law of the Koran.

Many people would be surprised to learn that non-theists comprise 20.7% of today's military, and 27.8% of those serving in the Guard and Reserve - and the percentage is increasing with every passing year. That is significant, but even so official military functions continue to include Christian prayers to the exclusion of Jews, Muslims, Buddhists, Zoroastrians, Wiccans - and atheists. It is time for Clebe McClary to realize he has no right to co-opt the acronym USMC to suit his own narrow interpretation of what is good and right in the world, because in doing so he is disrespecting all of the fine men and women who have worn our uniform proudly but do not share his point of view.

In declaring that Americans are free to practice their *chosen* religion, the Constitution also guarantees Americans the right to be free *from* religion if they so choose. Isn't that part of what we are fighting for?

While most soldiers, sailors, airmen and Marines practicing a religion have chaplains advocating for them, until recently all others have had to go it alone. The Military Association of Atheists and Freethinkers (www.militaryatheists.org) is an independent 501(c)3 organization which connects military members from around the world, and The Military Religious Freedom Foundation (www.militaryreligiousfreedom.org) is dedicated to ensuring all members of the Armed Forces fully receive the Constitutional guarantee of religious freedom to which they and all Americans are entitled.

# ODD OOD

*Most likely only a Marine, or someone who has read "Salty Language," can decipher the title of this story. It was told by Dr. Albert C. Pierce, the Director of the Center for the Study of Professional Military Ethics at The United States Naval Academy, as he was introducing General James Mattis, who was there to give a lecture on Ethical Challenges in Contemporary Conflict in the spring of 2006. This was taken from the transcript of that lecture.*

A couple of months ago, when I told General Krulak, the former Commandant of the Marine Corps, that we were having General Mattis speak this evening, he said, "Let me tell you a Jim Mattis story."

General Krulak said when he was Commandant of the Marine Corps, every year - starting about a week before Christmas - he and his wife would bake hundreds and hundreds and hundreds of Christmas cookies. They would package them in small bundles. Then on Christmas day, he would load his vehicle. At about four AM, General Krulak would drive himself to every Marine guard post in the Washington-Annapolis-Baltimore area and deliver a small package of Christmas cookies to whatever Marines were pulling guard duty that day.

He said that one year, he had gone down to Quantico as one of his stops to deliver Christmas cookies to the Marines on guard duty. He went to the command center and gave a package to the lance corporal who was on duty.

He asked, "Who's the Officer of the Day?" The lance corporal said, "Sir, it's Brigadier General Mattis." And General Krulak said, "No, no, no. I know who General Mattis is. I mean, who's the Officer of the Day today, Christmas Day?" The lance corporal, feeling a little anxious, said, "Sir, it is Brigadier General Mattis."

General Krulak said that, about that time, he spotted in the back room a cot, or a daybed. He said, "No, Lance Corporal. Who slept in that bed last night?" The lance corporal said, "Sir, it was Brigadier General Mattis." About that time, General Krulak said that General Mattis came in, in a duty uniform with a sword, and General Krulak said, "Jim, what are you doing here on Christmas Day? Why do you have duty?"

General Mattis told him that the young officer who was scheduled to have duty on Christmas had a family, and General Mattis decided it was better for the young officer to spend Christmas Day with his family, and so he chose to take the duty himself.

General Krulak said, "That's the kind of officer that Jim Mattis is."

# LIFE IS LIKE...
## *A Case of C Rations*

**Major General Robert Scales**

*I have taken the liberty of adapting these remarks for Marines... since all Vietnam vets - and all combat vets for that matter - can identify with the General's observations.*

I'm proud I served in Vietnam. Like you, I didn't kill innocents - I killed the enemy. I didn't fight for big oil or for some lame conspiracy - I fought for a country I believed in and for the buddies who kept me alive. Like you I was troubled that, unlike my father, I didn't come back to a grateful nation. It took a generation and another war, Desert Storm, for the nation to come back to me. Also like you I remember the war being ninety-nine percent boredom and one percent pure abject terror - but not all my memories of Vietnam are terrible. There were times when I enjoyed my service in combat. Such sentiment must seem strange to a society today that has, thanks to our superb volunteer military, been completely insulated from war. If they thought about Vietnam at all our fellow citizens would imagine that fifty years would have been sufficient to erase this unpleasant war from our conscientiousness. Looking over this assembly it's obvious that the memory lingers, and those of us who fought in that war remember.

327

The question is, why? If this war was so terrible why are we here? It's my privilege today to try to answer that question not only for you, brother veterans, but maybe for a wider audience for whom, fifty years on, Vietnam is as strangely distant as World War One was to our generation. Vietnam is seared in our memory for the same reason that wars have lingered in the minds of soldiers for as long as wars have been fought. From Marathon to Mosul young men and now women have marched off to war to learn that the cold fear of violent death and the prospects of killing another human being heighten the senses and sear these experiences deeply and irrevocably into our souls and linger in the back recesses of our minds. After Vietnam we may have gone on to thrilling lives or dull - we might have found love or loneliness, success or failure - but our experiences have stayed with us in brilliant Technicolor and with a clarity undiminished by time. For whatever primal reason, war heightens the senses. When in combat we see sharper, hear more clearly and develop a sixth sense about everything around us.

Remember the sights? I recall sitting in the jungle one bright moonlit night marveling on the beauty of Vietnam. How lush and green it was. How attractive and gentle the people, how stoic and unmoved they were amid the chaos that surrounded them.

Do you remember the sounds? Where else could you stand outside a bunker and listen to the cacophonous mix of Jimmy Hendrix, Merle Haggard and Jefferson Airplane? Or how about the sounds of incoming? Remember it wasn't a boom like in the movies, but a horrifying noise like a passing train followed by a crack and the whistle of flying fragments.

Remember the smells? The sharpness of cordite, the choking stench of rotting jungle and the tragic sweet smell of enemy dead.

I remember the touch, the wet, sticky sensation when I touched one of my wounded Marines one last time before the medevac rushed him forever from our presence but not from my memory, and the guilt I felt realizing that his pain was caused by my inattention and my lack of experience. Even taste is a sense that brings back memories. Remember the end of the day after the log bird flew away leaving mail, C rations and warm beer?

Only the first sergeant had sufficient gravitas to be allowed to turn the C ration cases over so that all of us could reach in and pull out a box on the unlabeled side - while hoping that it wasn't going to be ham and lima beans again.

Look, forty years on I can forgive the guy who put powder in our ammunition so foul that it caused our M-16s to jam. I'm okay with helicopters that arrived late. I'm over artillery landing too close and the occasional canceled air strike. But I will never forgive the Pentagon bureaucrat who in an incredibly lame moment thought that a Marine or soldier would open a can of that green, greasy, gelatinous goo called ham and lima beans and actually eat it.

But to paraphrase that iconic war hero of our generation, Forrest Gump, life is like a case of C Rations - you never know what you're going to get - because for every box of ham and lima beans there was that rapturous moment when you would turn over the box and discover the bacchanalian joy of peaches and pound cake. It's all a metaphor for the surreal nature of that war and its small pleasures... those who have never known war cannot believe that anyone can find joy in hot beer and cold pound cake. But we can.

Another reason why Vietnam remains in our consciousness is that the experience has made us better. Don't get me wrong. I'm not arguing for war as a self improvement course. And I realize that war's trauma has damaged many of our fellow veterans physically, psychologically and morally. But recent research on Post Traumatic Stress Disorder by behavioral scientists has unearthed a phenomenon familiar to most veterans - that the trauma of war strengthens, rather than weakens, us (they call it Post Traumatic Growth). We know that a near death experience makes us better leaders by increasing our self reliance, resilience, self image, confidence and ability to deal with adversity. Combat veterans tend to approach the future wiser, more spiritual and content with an amplified appreciation for life. We know this is true. It's nice to see that the human scientists now agree.

I'm proud that our service left a legacy that has made today's military better. Sadly Americans too often prefer to fight wars with technology. Our experience in Vietnam taught the nation the lesson that war is inherently a human, rather than a technological, endeavor. Our experience is a distant whisper in the ear of today's technology wizards that firepower is not sufficient to win, that the enemy has a vote, that the object of war should not be to kill the enemy but to win the trust and allegiance of the people and that the ultimate weapons in this kind or war are superbly trained, motivated, and equipped troops who are tightly bonded to their buddies and who trusts their leaders. I've visited our young men and women in Iraq and Afghanistan several times, and on each visit I've seen firsthand the strong connection between our war and theirs. These are worthy warriors who operate in a manner remarkably reminiscent of the way we fought so many years ago. The similarities are surreal.

Close your eyes for a moment and it all comes rushing back. In Afghanistan I watched Marines from my old unit as they conducted daily patrols from firebases constructed and manned in a manner virtually the same as those we occupied and fought from so many years ago. Every day these devil dogs trudge outside the wire and climb across impossible terrain with the purpose, as one sergeant put it - to "kill the bad guys, protect the good guys and bring home as many of my Marines as I can." Your legacy is alive and well. You should be proud.

The timeless connection between our generation and theirs can be seen in the unity and fighting spirit of our troops in Iraq and Afghanistan. Again and again, I get asked the same old question from folks who watch them in action on television. Why is their morale so high? Don't they know the American people are getting fed up with these wars? Don't they know Afghanistan is going badly? Often they come to me incredulous about what they perceive as a misspent sense of patriotism and loyalty.

# Been There, Done That... Got the T-Shirt!

I tell them time and again what every one of you sitting here today, those of you who have seen the face of war, understand. It's not really about loyalty. It's not about a belief in some abstract notion concerning war aims or national strategy. It's not even about winning or losing. On those lonely firebases as we dug through C ration boxes and drank hot beer we didn't argue the righteousness of our cause or ponder the latest pronouncements from McNamara or Nixon or Ho Chi Minh for that matter. Some of us might have trusted our leaders, or maybe not. We might have been well informed and passionate about the protests at home, or maybe not. We might have groused about the rich and privileged who found a way to avoid service, but we probably didn't. We might have volunteered for the war to stop the spread of global communism, or maybe we just had a failing semester and got swept up in the draft.

In war young Marines think about their buddies. They talk about families, wives and girlfriends and relate to each other through very personal confessions. For the most part the military we served with in Vietnam did not come from the social elite. We didn't have Harvard degrees or the pedigree of political bluebloods. We were in large measure volunteers and draftees from middle and lower class America. Just as in Iraq today we came from every corner of our country to meet in a beautiful yet harsh and forbidding place, a place that we've seen and experienced but can never explain adequately to those who were never there.

Marines suffer, fight and occasionally die for each other. It's as simple as that. What brought us to fight in the jungle was no different than the motive force that compels young Marines today to kick open a door in Ramadi with the expectation that what lies on the other side is either an innocent huddling with a child in her arms or a fanatic insurgent yearning to buy his ticket to eternity by killing the infidel. No difference. Patriotism and a paycheck may get a Marine into the military, but fear of letting his buddies down gets him to do things that might just as well get him killed.

What makes a person successful in America today is a far cry from what would have made him a success in the minds of those assembled here today. Big bucks gained in law or real estate, or big deals closed on the stock market made some of our countrymen rich - but as they have grown older they now realize that they have no buddies. There is no one who they are willing to die for, or who is willing to die for them. William Manchester served as a Marine in the Pacific during World War II and put the sentiment precisely right when he wrote, "Any man in combat who lacks comrades who will die for him, or for whom he is willing to die, is not a man at all. He is truly damned."

The Anglo Saxon heritage of buddy loyalty is long and frightfully won. Almost six hundred years ago the English king, Henry V, waited on a cold and muddy battlefield to face a French army many times his size. Shakespeare captured the ethos of that moment in his play *Henry V*. To be sure Shakespeare wasn't there, but he was there in spirit because he understood the emotions that gripped and the bonds that brought together both king and soldier. Henry didn't talk about national strategy. He didn't try to justify the faulty intelligence or ill formed command decisions that put his soldiers at such a terrible disadvantage. Instead, he talked about what made English soldiers fight and what in all probably would allow them to prevail the next day against terrible odds. Remember this is a monarch talking to his men:

*"This story shall the good man teach his son; And Crispin Crispian shall ne'er go by, From this day to the ending of the world, But we in it shall be remembered - We few, we happy few, we band of brothers; For he to-day that sheds his blood with me Shall be my brother; be he ne'er so vile, This day shall gentle his condition; And gentlemen in England now-a-bed Shall think themselves accurs'd they were not here, And hold their manhoods cheap whiles any speaks That fought with us upon Saint Crispin's day."*

You all here assembled inherit the spirit of St. Crispin's day. You know and understand the strength of comfort that those whom you protect, those in America now abed, will never know. You have lived a life of self awareness and personal satisfaction that those who watched you from afar in this country, who hold their manhood cheap, can only envy.

I don't care whether America honors or even remembers the good service we performed in Vietnam. It doesn't bother me that war is an image that America would rather ignore. It's enough for me to have had the privilege to be among you. It's sufficient to talk to each of you about things we have seen and kinships we have shared in the tough and heartless crucible of war.

Someday we will all join those who are serving so gallantly now and have preceded us on battlefields from Belleau Wood to Fallujah. We will gather inside a firebase to open a case of C rations with *every* box peaches and pound cake. We will join with a band of brothers to recount the experience of serving something greater than ourselves. I believe in my very soul that the almightily reserves a corner of heaven, probably around a perpetual campfire, where someday we can meet and embrace all of the band of brothers throughout the ages to tell our stories while envious standers-by watch and wonder how horrific and incendiary the crucible of violence must have been to bring such a disparate assemblage so close to the hand of God.

**Adapted from a speech by Major General Robert Scales USA (Ret) at the Truman Library on 12 September 2009.**

# SHOULD WE ASK, SHOULD THEY TELL?

*Since I wrote this in late 2009 Congress has gone ahead and gotten rid of the "Don't Ask, Don't Tell" provision which prohibits gays from serving openly in the Armed Forces - but even so, my point remains valid.*

According to the latest news cycle Senate Majority Leader Harry Reid is planning to schedule a vote targeting the repeal of the Pentagon's "Don't Ask, Don't Tell" policy immediately after Thanksgiving. In my opinion that is a big mistake - if by any chance you care about our military's ability to defend this nation.

First, before I get excoriated by progressives and labeled a "hater," let me explain where I stand and why I feel that way - and I promise I will do it without making any moralistic or religious arguments. I will leave that to others.

So, do I think "Lesbians, Gays, Bi-Sexuals and Transgenders" (LGBTs) are normal? Well, from a strictly practical standpoint - no. Think about it. If all of the animals going two by two into Noah's Ark (or all of the organisms evolving from cosmic soup, if you prefer) had been gay, there would be no lions, tigers or bears in the world today. Going back a bit further, how far would the human race have gotten if the Garden of Eden had been initially populated by, say, Adam and *Steve?* The fact of the matter is gays do not procreate, and when you get right down to it isn't that the whole purpose of sex? And remember, that holds true whether you believe in evolution or creation.

Okay, now that I've addressed the normalcy issue, let's move on to my personal feelings about our LGBT community. I think it can be best summed up by something I remember seeing written on a sidewalk in Australia - "We're here, we're queer, and we're not going away." That is obviously a true statement on all three counts. LGBTs are human beings with a right to life, liberty and the pursuit of happiness like everyone else. I have no problem with people pursuing an "alternate lifestyle" while holding down a job as a cook at McDonalds or executive with IBM or pilot for United Airlines. For that matter, why not make one the head of the Department of Homeland Security, or even a justice on the Supreme Court! If they had been denied these positions based solely upon their sexuality that would have been wrong, and I would be among the first to say it.

So why, if I feel that way, do I oppose repeal of DADT? It's quite simple. The cook, executive, pilot, bureaucrat and judge go home at the end of the day to whatever significant other they may have. Those of us in the military do not! We must share rooms, tents, showers and sometimes blood with one another, and the need for unit cohesion is very real.

Proponents of repealing DADT will tell you that denying gays access to the military lifestyle will somehow infringe upon their ability to "pursue happiness." What? Haven't these people ever seen *Private Benjamin?* How long will it be before they start asking, "Where's the condos?" Now, those who truly want to serve, in contrast to the majority who simply want to make a political statement, should be applauded for their patriotism - but if they really want to make our Nation safer they should instead take one for the team and opt out.

This has nothing to do with discrimination. Champions of political correctness may be surprised to learn military recruiters can't sign up Americans in wheelchairs. If you have AIDS or other illnesses, you are also excluded. If you score too low on the ASVAB test, aren't tall enough, weigh too much, or for that matter have flat feet you will be turned away. The military even prohibits the enlistment of communists - while there is apparently no such prohibition in place for members of Congress, or an occupant of the White House. Service in the military is not an inalienable right. It does not matter how patriotic the applicant is. It doesn't matter if they feel their "pursuit of happiness" has been derailed. They bottom line is some profiles and lifestyles are incompatible with military service.

This discussion should also be broken down into its four elements, since the prohibition on "gays" in the military is too general and doesn't tell the whole story - and I have heard very little dialog about the other three. Let's address them in order.

Lesbians, from what I understand, come in two very different varieties - butch and femme. Will butch female Marines become known as "Devil Dykes"? I don't know, but a recent comment by comedian Dennis Miller gave a good indication when he said he'd like to see DHS Secretary Janet Napolitano go through a body scanner because he has a bet riding on the outcome. I'd say it's even money.

Gays are another matter altogether. I just can't see them, with their flamboyant sense of style, voluntarily donning olive drab - and it won't be long before there is a push to get that changed to the pink camouflage outfits which are all the rage down at the mall. Stylish to be sure, but not very effective in combat.

Bisexuals, for their part, are unsuitable for any sort of leadership positions. After all, one of the fourteen leadership traits is decisiveness, and these people can't even decide which "team" they want to play on!

Which brings us to transgenders. I don't even know where to start with this one. Will they be considered male, or female? Which head will they use? Which physical fitness test will they run? And if they change genders while serving, will it be standard policy to change the basic uniform issue and swap jock straps for underwire brassieres? Talk about confusing!

The preceding remarks may seem a bit flip, but from a practical standpoint - how do we billet these newly assimilated individuals? Do we put gays and straights together in a room to cut down on the hanky panky (and in doing so, probably violate the *strait* member's rights and comfort level), do we billet "birds of a feather" in the same nest, or should we make it a random process? Let's look at this question a different way. Women have been a part of the military for quite some time now and have been assimilated into all occupational fields save those involving direct combat, but even so no one has ever suggested putting them in the same barracks room with men. Could there be a reason for this? And how would a straight sharing a bedroom and shower with a gay be different? Naturally the gay community will say, "That's easy, give everyone their own room!" - but does it really make sense to restructure an entire organization and spend billions of dollars we don't have to construct new barracks in order to make a point... and in any case there are no rooms out in the field.

It has to be asked - where does this "fairness doctrine" stop? Will dogtags now include sexual preference along with religion, blood type and gas mask size? And don't forget, necrophiliacs and practitioners of bestiality are people too. They are also Americans with the right to life, liberty and the pursuit of happiness - in whatever bizarre form that may be - aren't they? Does that mean we must make a place for them at the table - or in the barracks room - as well? Aren't *their* rights being violated? Or, as a practical matter, how would you feel about it if the mortician attending to your favorite Uncle Tom's remains was a necrophiliac, or if a devotee of bestiality took your beloved pet schnauzer out for a walk?

Admiral Mullen, by coming out in favor of repealing DADT, has unwittingly confirmed something Marines have been teasing sailors about for centuries. I am reminded of the scene in *A Few Good Men* where Colonel Nathan Jessup, portrayed so ably by Jack Nicholson, told Tom Cruise to "stand there and that faggoty white uniform and extend me some respect." We've always had our suspicions... so how about a compromise, Admiral Mullen? Rather than jump in with both feet, let's try this incrementally. Since you are so gung ho about supporting the LBGT agenda, let's make the Navy a test case so the other services can see how things work out for you. Go ahead, lead by example!

General Amos and his predecessor as Marine Commandant, General Conway, made it abundantly clear that the Marine Corps doesn't want to participate in this game of politically correct footsie. So why force the issue? Why does it have to be ALL of the services? There's an old saying which says if something is working properly, don't tinker with it, and by any standard of measurement the Marine Corps has been working just fine for the past two hundred and thirty-five years.

One last thought. Before taking a vote I suggest Barack Obama, Harry Reid and any other politician championing the repeal of DADT (especially those who have never served in uniform, like those two) spend a few days billeted in a Congressional barracks room sharing a shower with Barney Frank. It could be a real eye opener.

# ABOUT THE AUTHOR

Andy Bufalo retired from the Marine Corps as a Master Sergeant in January of 2000 after more than twenty-five years service. A communicator by trade, he spent most of his career in Reconnaissance and Force Reconnaissance units but also spent time with Amtracs, Combat Engineers, a reserve infantry battalion, and commanded MSG Detachments in the Congo and Australia.

He shares the view of Major Gene Duncan, who once wrote "I'd rather be a Marine private than a civilian executive." Since he is neither, he has taken to writing about the Corps he loves.

*Semper Fi!*